REEDS INTRODUCTIONS

Essential Sensing and Telecommunications
For Marine Engineering Applications

REEDS INTRODUCTIONS

Essential Sensing and Telecommunications for Marine Engineering Applications

Physics Wave Concepts for Marine Engineering Applications

REEDS INTRODUCTIONS

Essential Sensing and Telecommunications
For Marine Engineering Applications

Christopher Lavers

ADLARD COLES NAUTICAL

BLOOMSBURY

LONDON · OXFORD · NEW YORK · NEW DELHI · SYDNEY

Thomas Reed
An imprint of Bloomsbury Publishing Plc

50 Bedford Square	1385 Broadway
London	New York
WC1B 3DP	NY 10018
UK	USA

www.bloomsbury.com

REEDS, ADLARD COLES NAUTICAL and the Buoy logo are trademarks of Bloomsbury Publishing Plc

First published 2017

British Library Cataloguing-in-Publication Data
A catalogue record for this book is available from the British Library.

Library of Congress Cataloguing-in-Publication data has been applied for.

ISBN: PB: 978-1-4729-2218-2
ePDF: 978-1-4729-2220-5
ePub: 978-1-4729-2219-9

2 4 6 8 10 9 7 5 3 1

Typeset in Myriad Pro by Newgen Knowledge Works (P) Ltd, Chennai, India
Printed and bound in Great Britain by CPI Group (UK) Ltd, Croydon CR0 4YY

To find out more about our authors and books visit www.bloomsbury.com. Here you will find extracts, author interviews, details of forthcoming events and the option to sign up for our newsletters.

Dedication

Isaiah 53: 5

ה וְהוּא מְחֹלָל מִפְּשָׁעֵנוּ, מְדֻכָּא מֵעֲוֹנֹתֵינוּ; מוּסַר שְׁלוֹמֵנוּ עָלָיו, וּבַחֲבֻרָתוֹ נִרְפָּא לָנוּ.

But he was wounded because of our transgressions, he was crushed because of our iniquities: the chastisement of our welfare was upon him, and with his stripes we were healed.

I thank my wife Anne for her patience and support, my family for 'lost time', Professor Alan Myers (Director of Studies at BRNC), and key imagery sources: James Burr and John Stevenson at Leonardo Airborne and Space Systems Division, Charlie Reed at FLIR Systems, Mark Bown at Kelvin Hughes, Cathy Millar at Wiska UK Marine Products, and Dr David Jones for help with flow monitoring systems. I would lastly like to thank Jenni Davis for her accurate checking of the completed manuscript.

Digital Sequence Sara-Kate Lavers © 2016

CONTENTS

6 Radar Echo Signal Processing in a Real Operational Environment   140

7 Electro-Optical and Thermal Systems   164

INTRODUCTION

This book is designed to support your learning of the essential Sensing and Telecommunications topics relevant to the marine environment. An understanding of wave motion concepts is needed to provide a firm foundation for these subjects, with a comprehensive study of the basic principles of wave motion covered in Reeds *Physics Wave Concepts for Marine Engineering Applications*.

Sensors and communications are the 'flip side' of the same coin of electromagnetic wave applications, as are frequency and wavelength. This book focuses on use of electromagnetic waves in the above water environment but discussion of sonar is given. We will look at key sensors and telecommunications maritime skills, relevant to the marine environment, underpinning vital International Maritime Organization (IMO) recommendations to support safety at sea. These topics are also important for developing competence to achieve the award of Navigational Watch Certificate (NWC) within the RN STCW II/1 Code.

This book is valuable to all students of maritime science, and is focused at the foundational level of study required by maritime users. It follows a similar pattern to others in the Reeds Marine Engineering and Technology Series, providing both key text and numerous worked examples.

By the end of the book you should have a clear understanding of the following topics:

Radar principles: Including relevant calculations throughout, basic radar signal processing, explaining the appropriate anti-noise or anti-clutter technique to be used.

Electro-optics principles: The requirements for imaging in low level light conditions or the absence of visible light entirely.

Sensors: How to select a particular sensor for different roles, platforms and environments. Common sensor types widely deployed across varied platforms globally include: radar, image intensifier or night vision electro-optical devices, thermal imagers, lasers, and laser range finders for maintaining distance while undergoing replenishment at sea, often with low light level adaptation fitted.

Communications: Readers will describe the key differences between wired and wireless methods, analogue and digital transmission, propagation, analogue to digital conversion, and encryption. There are many forms of maritime communication with diverse aims and objectives regarding their communications requirement. In short, communications has a vital role in ship safety at sea and in providing up-to-date, real-time situational analysis. There is often a requirement to transfer significant amounts of data to other maritime users in near real time. Hence, significant aspects of the design are focused upon communications and data-link systems.

Roles: The role of a platform has considerable bearing on which sensor types or 'fits' are installed. In some cases, platforms are designed to carry out a single role; however, given the costs involved today in modern platforms (manned or unmanned) and to enable maximum flexibility, most platforms – ship, satellite or otherwise – fulfil several functions.

The style of writing this book considers the common ground encountered by all maritime users and will thus be focused on the common ground of 'peaceful' applications for the modern seafarer!

Dr Christopher Lavers
PhD (Exon), BSc (Exon), PGCE (LTHE), M. Inst. P, C. Phys., FHEA

1

Basic Principles of Communications

'Lion Switching: The whole lion from user to user must be set up before information is sent to the transmitter.'

Poor exam answer (an extremely dangerous form
of switching CRL)

1.1 The purpose of maritime communications

Human communication involves exchanges of information in various ways. However, it is the more specialised use of communications for maritime users that we will consider here. Maritime communications must provide a practical way of communicating with shore authorities, other ships, aircraft, satellite and remotely controlled UAVs, worldwide, day or night. There are different forms of maritime communication and many different organisations with different aims and objectives regarding their communications requirements, e.g. coastguard, merchant and naval vessels, or indeed private pleasure craft, yachts and super-yachts. Clearly correct terminology and understanding of language, as illustrated by the quote above, is vital.

1.2 History of key telecommunications maritime-related advances

There have been many significant developments in communications throughout history, up to the modern telecommunications era. The highlights of which are listed below:

The first developments were predominantly visual, auditory and non-electrical methods:

In antiquity: Visual (fires: beacons and smoke signals), audible (drums and horns).

6th century BC: Mail (attributed to the Persian Cyrus the Great, 550 BC).

5th century BC: Pigeon post, the first 'air mail' (attributed to the Persians and used by Julius Caesar).

4th century BC: Greek hydraulic semaphore (this system involved identical containers of water on two hills, each with a vertical rod floating in it; the rods were inscribed with predetermined codes at various points along their height).

15th century AD: Maritime flag semaphores, and the first 'flaggers'.

1672: The first experimental acoustic (mechanical) telephone, with ranges of up to 800 m under ideal conditions [1.1].

1792: Semaphore lines (or optical telegraph, developed first by Frenchman Claude Chappe).

1867: Signal lamps (e.g. flashing dots and dashes from lanterns, first demonstrated by Vice Admiral Colomb RN) lead to the Aldis lamp and heliograph.

(i) *Basic electrical signals:*
1837: First commercial electric telegraph (demonstrated by Cooke and Wheatstone, although the most universally accepted of the 60 or more different telegraphy methods proposed was that of Samuel Morse).

1858: First transatlantic electrical telegraph cable (16 August).

1876: Telephone (invented by Alexander Graham Bell).

(ii) *Advanced electrical and electronic signals:*
1893: Wireless telegraphy.

1896: Radio.

1914: First North American transcontinental telephone call (the 'audion').

1927: Television (Ives and Gray, Bell Telephone Laboratories, USA.)

1936: World's first public videophone network (Schubert, Germany).

1946: Limited capacity mobile telephone service for automobiles (Bell Systems, USA).

1956: Transatlantic telephone cable (TAT-1 Oban, Scotland to Clarenville, Newfoundland).

1960: Echo – passive satellite platform.

1962: Commercial telecommunications satellite (Telstar 1). First active repeater or transponder.

1964: Fibre optic telecommunications (Kao and Hockham, Standard Telephones and Cables (STC), UK).

1969: Computer networking.

1973: The first handheld mobile phone (demonstrated by John Mitchell and Martin Cooper, Motorola).

1981: First mobile (cellular) phone network.

1982: Inmarsat first ship-to-shore satellite communications operations.

1982: Simple Mail Transfer Protocol (SMTP) email.

1983: Internet.

1999: First limited mobile satellite handheld phones (Globalstar and Iridium).

2003: Voice over Internet Protocol (VoIP) Internet telephony.

1.2.1 Maritime signalling

Before 1860, signalling was mostly performed by means such as sail movements, smoke, gun firing and flags. In this period, the carrier pigeon provided the first rapid long-distance communications. A chain of relay stations was frequently set on mountain peaks and fires were used for signalling information, notably the beacons established to signal the approach of the Spanish Armada to the English fleet (1588), and similar to those seen in modern films, e.g. the lighting of the beacons of

Gondor in *The Lord of the Rings: The Return of the King*. During the period 1780–90 a flag code system was developed, and later a further system of hoisting pennants and flags. In 1795 the semaphore signalling method was adopted, allowing a message to be sent by means of visual signalling using two flags, either manually or by a mechanical device with two moving arms. Coloured lights as used in maritime 'rules of the road', and pyrotechnic coloured smoke and flare signals are still used for distress and emergency purposes.

Signalling between ships with flags and semaphore was used for hundreds of years, but these methods suffered from a lack of speed and short Line Of Sight (LOS) range. The electric telegraph (1837) was the first step towards a modern communication system, although initially limited by being a wired-line system requiring the development of subsea cables to span the oceans. This system transmitted and received electrical impulses. Samuel Morse introduced the Morse code, a system of dots and dashes in different arrangements to represent letters and numbers.

In the late 19th century, the remote Cornish village of Porthcurno became inadvertently famous as the British terminal of early submarine telegraph cables, the first landed in 1870 as part of an international link stretching from the UK to India. Porthcurno was a last-minute choice because of the risk of damage to the cables by ships' anchors in the River Fal estuary en route to Falmouth, the planned destination. In 1872, the Eastern Telegraph Company (ETC) Limited was formed. ETC and its cable operations expanded into the early 20th century, merging with Marconi's Wireless Telegraph Company Limited in 1928 to form Imperial and International Communications Limited, renamed Cable and Wireless Limited in 1934. Between the wars the Porthcurno cable office operated 14 cables, for a time becoming the largest submarine cable station in the world, able to receive and transmit 2 million words a day. Further Porthcurno details are at telegraphmuseum.org.

In 1865 flashing lights were introduced as signalling devices, using Morse code. Unlike flag signalling, the alphabet and numbers are written in an arrangement or combination of dots and dashes signalled by flashing a light on and off for a designated period of time to represent a dot or a dash. Scottish mathematician James Clerk Maxwell (1831–1879) first predicted the existence of electromagnetic waves in 1864, but it wasn't until 1887 that German physicist Heinrich Hertz (1857–1894) used a spark-gap device to produce and detect such waves in the radio spectrum. Guglielmo Marconi (1874–1937) first recognised its potential for practical applications with his own radio experiments, opening the field of communication engineering and maritime communications. Born in Bologna, Italy, to an Irish mother and Italian father, a landowner, Marconi's defining moment was while on holiday in the Italian Alps in 1894 when he read a paper by Heinrich Hertz describing his experiments. From that moment, the subject became Marconi's lifelong obsession. Within a year he had transmitted Morse signals about a kilometre in distance, to positions beyond the crest of a nearby mountain.

While Marconi was conducting his experiments, Admiral of the Fleet Sir Henry Jackson (1855–1929) was undertaking similar work in Devonport, England, achieving the first ship-to-ship wireless communications and communication with vessels up to 3 miles away, using a small group of vessels forming the HMS *Defiance* Torpedo School, off Wearde Quay, Saltash [1.2]. Marconi's equipment, however, was selected for Royal Navy fleet trials in 1899, achieving ranges of 87 miles between ships. Admiral Jackson later worked with Marconi to develop a fleet wireless system and Jackson's achievement was recognised with election as a Fellow of the Royal Society in 1901 [1.3].

The British Admiralty ordered a ship-fitting programme and thus Royal Naval radio telecommunications began. The first telecommunication systems were telegraphs using simple spark-gap transmitters with an induction coil, more sophisticated than Hertz's original device. John Ambrose Fleming's invention of the thermionic valve in 1904 permitted greater control of transmission and enabled the first voice transmission.

Communications has now grown into a worldwide technology and industry using sophisticated techniques, such as internet and satellite systems, to deliver information anywhere on land, sea or in the air, in various forms such as data, text, voice, and still and moving pictures. The modern requirements for a ship's internal communications are also widespread, from telephone and message requirements, intercom and announcement systems to internet access.

1.3 The communications channel

When people exchange information, they use a communications channel. This basically consists of a transmitting and a receiving device linked by a bearer of electromagnetic waves (line or other medium), whether we consider two people in a face-to-face conversation or two people conducting point-to-point communication at long range using the interface of telephone communications equipment. A block diagram of a communications channel is shown in figure 1.1.

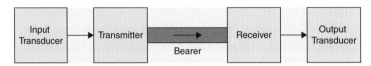

Figure 1.1: *Simple block communications channel diagram.*

We consider communications channels in terms of availability, speed, information carrying capacity and noise immunity. Channels are also commonly defined as simplex, half-duplex, and (full) duplex, as shown in table 1.1.

Channel type	Characteristics
Simplex	The simplest communications channel, capable of one-way communication only. A simplex circuit requires only one transmitter, receiver and bearer, but can be used with multiple receivers as in TV, radio and ship-wide broadcasts.
Half-duplex	Capable of two-way communication, but only one way at a time. A half-duplex circuit requires two transmitters and two receivers, but only one bearer.
Full duplex	Capable of simultaneous two-way communication. A full duplex circuit generally requires two transmitters, two receivers and two bearers (equivalent to two separate simplex circuits).

Table 1.1: *Different channel types and their characteristics.*

1.3.1 Input and output transducers

The information to be sent, transmitted through modern telecommunications systems, can come from many sources, e.g. human voices, music (continuous waves). These waves are in a form that can be easily interpreted by human senses, principally sight and hearing. However, electronic communications systems require information to be sent as an electrical signal, with information converted into an electrical signal prior to transmission. In the receiver, detected electrical signals are converted back into a form a human can recognise. Devices that perform these energy conversions are known as transducers and the electrical form of the information is called the electrical baseband.

A transducer is simply any device that converts energy in one form into energy in another form. In communications systems, the input transducer is used to convert the signal information into the electrical baseband signal, while the output transducer converts the electrical baseband signal back into a form a person will understand. The more common input and output transducer paired combinations are shown in table 1.2.

There is a requirement to match input and output transducers to the relevant information, e.g. microphone and loudspeaker for voice communications. However, modern sonar applications usually take received acoustic sonar energy and display returning frequencies on a video screen in frequency bands. Communications systems contain other transducers, the most common of which is the piezoelectric device. In some solid materials, such as crystals and ceramics, electrical charge can be generated in direct response to applied mechanical stress such as squeezing or pressing. The piezoelectric effect is a reversible process, as is the reverse piezoelectric effect (internal generation of a mechanical strain resulting from an applied electric field). Piezoelectricity is used in sound detection and transmission, and high voltage generation. These devices convert pressure in the form of dimensional shape changes of the device into electrical energy and vice versa.

Transducer	Transducer type	Energy input	Energy output
Microphone	Input	Sound	Electrical
Loudspeaker	Output	Electrical	Sound
Camera (video)	Input	Moving or still pictures	Electrical
Monitor (TV or computer)	Output	Electrical	Moving or still pictures
Photodiode	Input	Light levels	Electrical
Light Emitting Diode (LED)	Input	Electrical	Light levels
Facsimile (transmit)	Input	Still picture	Electrical
Facsimile (receive)	Output	Electrical	Still picture

Table 1.2: *Input and output transducers.*

1.3.2 Transmitters and receivers

A transmitter converts the electrical baseband into a form that the bearer can carry, e.g. an electromagnetic wave for radio communication. Similarly, the receiver detects the signal and then converts it into an electrical baseband used to drive the output transducer.

1.4 Bearers and typical communications devices

Different bearers are used today for different electromagnetic wave bands. Copper metal cables have long carried electrical signals since their introduction in the mid-19th century. Copper is a very good metal to use as it is relatively common, and is drawn into wires easily. Copper has low electrical resistance (good conductivity) yet it is flexible so it can be bent without breaking, and it is cheaper than gold, which is preferred for low-noise and specialist applications for maritime forces. Low resistance means copper has low attenuation on key speech electrical baseband frequencies.

1.4.1 Twisted pair

The simplest cable is the twisted pair, consisting of two insulated copper trunk cables to stop crosstalk between adjacent cables (figure 1.2). To reduce crosstalk further, and prevent interference, twisted pairs are surrounded by a conducting screen, creating a screened pair. Screened pairs can transmit frequencies up to a few kHz as shown by the graph in figure 1.2. The twisted pair wire is still the most widely used medium for telecommunications, although greater use is now made of broadband optical fibre. Telephone wires consist of two insulated copper wires twisted into pairs. Computer network wires consist of four pairs of copper wires used for voice and data transmission.

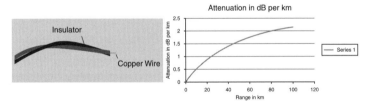

Figure 1.2: *Twisted or screened pair and attenuation in dB per km.*

In most twisted-pair ethernet configurations, repeaters are needed beyond 100 metres. Repeaters take time to regenerate the signal. This introduces propagation delays that affect network performance. As a result, network architectures limit the number of repeaters used.

The four primary electrical characteristics of twisted-pair cable are its series resistance R, series inductance L, shunt capacitance C and shunt conductance G per unit length of cable. Transmission characteristics, such as impedance Z and propagation constant γ, are calculated from these primary electrical characteristics.

Impedance is given by:

$$Z = \sqrt{\frac{R + j\omega L}{G + j\omega C}}$$

(**eq 1.1**)

For polyethylene cable, conductance G is small, so at low frequencies where $<< \dfrac{R}{L}$:

$$Z = \sqrt{\dfrac{R}{G + j\omega\, C}} \qquad \textbf{(eq 1.2)}$$

while at high frequencies, where $\omega >> \dfrac{R}{L}$:

$$Z = \sqrt{\dfrac{L}{C}} \qquad \textbf{(eq 1.3)}$$

Example 1.1: A twisted pair has a resistance of R = 50 ohms, a series inductance L = 4mH, R = 50 Ω, shunt capacitance C = 50 mF and minimal shunt conductance G per unit length of cable. What is the characteristic impedance Z for the twisted pair if the impedance is considered for high frequencies (2 decimal places)?

Using the equation for high frequencies where $\omega >> \dfrac{R}{L}$:

$$Z = \sqrt{\dfrac{4mH}{50mF}} = \sqrt{\dfrac{4}{50}} = 0.28 \text{ ohms (2 decimal places)}.$$

Twisted-pair cables have a long history in Voice Frequency (VF) transmission. Twisted-pair cable has been used for digital transmission since the 1960s. In North America, T carrier systems operate over twisted-pair cable at rates of 1.544 Mb/s (T1), 3.152 Mb/s (T1C) and 6.312 Mb/s (T2).

For maritime voice communications, it is important to observe basic protocols for radio communications. The IMO provides a standard ship reporting system for spelling out call signs, service abbreviations and words with the phonetic alphabet. Further details about IMO standard spelling is given in the UKHO *Admiralty List of Radio Signals* [1.4]. UKHO provides a comprehensive list of radio signals for pilot services, Vessel Traffic Services (VTS) and port operations worldwide, as well as meteorological observation station information.

1.4.2 Coaxial Cables

To extend frequency range further the capacitance needs to be reduced more than is achievable with twisted cables. Instead coaxial cable (figure 1.3) is used, having a central conductor surrounded by a low dielectric medium such as polythene or air and then a further earthed outer conducting sleeve, and finally an insulator. Use of a low dielectric medium reduces capacitance between a conductor and the sleeve, and extends the frequency range. Polythene dielectric coaxial cable is used up to a few hundred MHz, and with air dielectric up to 3 GHz. Coaxial cable is used for cable television systems, offices and Local Area Networks (LAN).

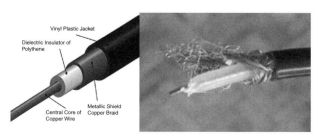

Figure 1.3: *Coaxial cable.*

1.4.3 Guided Electromagnetic Waves

Above 3 GHz, attenuation rises to unacceptable levels so other forms of transmission line must be used. Electromagnetic waves need to be constrained to follow the required path so intensity isn't reduced by free space spreading (inverse square law), instead attenuating more slowly with distance. This is achieved using waveguides, which are an efficient transmission method.

1.4.4 Waveguides

Waveguides are metallic tubes, of rectangular or circular cross section, used to carry waves at frequencies greater than 2 GHz (figure 1.4). To achieve low attenuation, it is usual to match the wavelength so the wall separation is just larger than half the wavelength, to guide the wave. Waveguides are viewed as a 'pipe' for waves, which are fed into one end of the 'pipe' travel along the waveguide and arrive with less attenuation than if they had been propagated into free space at the other end. Waveguides generally become too large below 2 GHz. For the interested reader, further details about waveguides are found elsewhere [1.5].

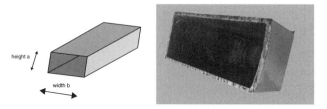

Figure 1.4: *Microwave waveguide (a) schematic and (b) picture of actual waveguide.*

Waveguides have large power-carrying capacities, limited by the breakdown potential of the air they contain. In radar, this capacity is increased by raising air pressure and by preventing moisture getting into the waveguide. It is common to pressurise waveguides, which increases the air's breakdown potential and prevents water ingress. Owing to their large power-carrying capacities, waveguides are used at frequencies down to 600 MHz for high power applications. In ships, waveguides are supplied in fixed lengths and corners assembled to make antenna feeders. Every joint and corner adds to overall waveguide transmission loss (signal attenuation) and ships are designed to minimise waveguide run lengths and the number of corners. To reduce the magnitude of such problems on antenna mountings, flexible waveguides have been developed, similar to flexible optical fibres for waveguide applications, consisting of a thin metallic tube surrounded by a thick rubber sheath (figure 1.5). Handled in long runs, it reduces the joint number and severity of bend required.

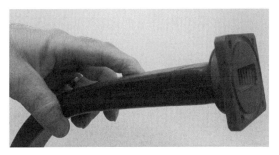

Figure 1.5: *Flexible microwave waveguide.*

1.4.5 Optical fibres

1.4.5.1 History of optical fibres

The most modern transmission line is the optical fibre. Surprisingly, as far back as Roman times, glass was drawn into fibres [1.6] but it was only in 1888, with Viennese doctors Roth and Reuss, that glass rods were first bent to illuminate body cavities. Seven years later, French engineer Henry Saint-Rene designed bent glass rods for guiding light in what we would now consider to be television. Heinrich Lamm first transmitted a bulb filament image through an optical fibre bundle in 1930, with the intention of looking inside inaccessible parts of the body, but the rise of Nazism forced Lamm, a German Jew, to move to America, abandoning his dream of becoming a medical professor.

In 1961, the American inventor Elias Snitzer published a theoretical description of single mode fibres whose core would be *so* small, just a few microns wide, that it could carry light in only one waveguide mode [1.7]. Snitzer demonstrated a laser directed through a thin glass fibre, sufficient for medical applications, however, light loss was too high for communication applications. Charles Kao and George Hockham published a theoretical paper in 1964 demonstrating that fibre light loss might be decreased by removing impurities such as copper and iron [1.8]. In 1970, scientists at Corning Glass Works made a single mode fibre with attenuation below 20 dB/km, achieved through doping silica glass with titanium.

In 1973, Bell Laboratories developed a Chemical Vapour Deposition (CVD) process that heated chemical vapours and oxygen to form ultra-transparent glass, mass produced into low-loss optical fibres. The first fibre-optic link was installed by the UK Dorset police in 1975. In 1977, the first live telephone traffic through fibre optics was achieved in Long Beach, California. In the late 1970s and early 1980s, telephone companies started to use fibres extensively.

In 1986, at the University of Southampton Professor David Payne developed the Erbium-Doped Fibre Amplifier (EDFA) [1.9], which reduced the cost of fibre systems greatly by eliminating optical-electrical-optical repeaters with the first all-laser-based amplifier (1988). Optical fibres doped with erbium and powered by tiny laser chips have revolutionised the way signals are regenerated for transcontinental communications and for fast data transmission over fibre-optic networks [1.10].

In 1987, Desurvire and Payne independently demonstrated that optical amplifiers could be built into the fibre-optic cable (figure 1.6). This all-optic system could carry 100 times more information than cables containing electronic amplifiers. In 1991, the photonic crystal fibre was developed by a colleague at the University of Southampton Philip St J Russell [1.11]. This type of fibre guides light via diffraction from periodic structures, which permits power to be carried more efficiently than conventional fibres, improving performance. The first all-optic fibre cable, TPC-5, using optical amplifiers, was laid across the Pacific in 1996. Today, various industries and maritime users employ fibre optic technology in many applications. A fibre optic transmission system has the same basic elements as a metal cable system – namely transmitter, cable bearer, repeaters and receiver – but with an optical fibre system the electrical signal is converted into light pulses.

Full exploitation of fibre systems coincided with development of the laser (Chapter 7). Optical fibre and lasers have provided the basis of modern optical fibre telecommunications. Pure glass is a bearer that carries information in short pulses of visible or infrared radiation. Optical fibres are now exclusively used with digital transmission systems. Optical fibres and Total Internal Reflection (TIR) propagation occurs between optically dense and optically less dense mediums, e.g. glass to air. There are three basic types of optical fibre: step index, graded index and single mode (or monomode).

In TIR, waves move from a 'slow' medium to a 'fast' medium and refract away from the normal. As the incident angle increases, both the reflected wave and the transmitted (refracted) wave increase in angle away from the normal, but there will eventually be an angle, the *critical angle*, where a refracted ray cannot enter the second medium because the transmitted (refracted wave) now runs along the boundary between the two media. Light striking the boundary at angles greater than the critical angle undergo TIR and remain in medium 1. The transmitted and reflected waves are two possible modes of the incident wave interacting with the surface, and when wave energy can no longer 'exit' in the transmitted wave it is instead 'channelled' into the remaining mode, which is the reflected wave.

Figure 1.6: *(a) Schematic optical fibre and (b) actual optical fibre bundle.*

Optical fibre is made of a central glass core (typically a refractive index n = 1.52) surrounded by cladding of slightly lower refractive index (typically n = 1.512) (figure 1.7a). The outer protective casing or 'jacket' is made of a strong opaque material. Core and cladding are made from a large rod and a tightly fitting cylinder drawn into a narrow fibre, encased in an outer jacket for protection. Laser pulses enter a fibre over a wide angle range, but only those that strike the boundary between core and cladding, at angles greater than the critical angle, propagate through the core. Waves at other angles are absorbed in the cladding and jacket and are lost.

Rays that propagate through a 'thick' core take different paths. Each path, or waveguide mode, is a slightly different length and takes a different time to travel the fibre's length (figure 1.7b). Paths 1 to 4 take respectively increased time to transit the same length of optical fibre.

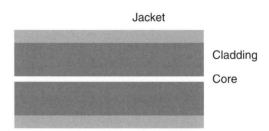

Figure 1.7a *Fibre optic structure.*

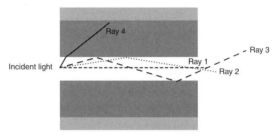

Figure 1.7b *Expanded optical fibre core view and 'multiple' guided modes.*

This range of transit times results in pulses spreading out or dispersing along the fibre, and the resultant spreading places a minimum time needed between pulses to resolve them separately to distinguish between the ones and zeros of a digital sequence transmitted. The more modes there are, the greater the spreading and thus the greater the time required between pulses for pulses to be resolved in the receiver (figure 1.8). Further theoretical treatment is found elsewhere [1.12].

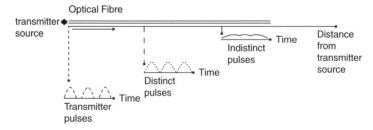

Figure 1.8: *Diagram showing multimode optical fibre intermodal dispersion.*

This spreading process is termed *intermodal dispersion* and in practice multimode fibres severely limit the distance possible and digital data rate used. The number of modes depends on the relationship between fibre diameter and laser wavelength. If the optical fibre diameter is large (greater than 10 wavelengths), there will be so many modes that the fibre is termed a **multimode fibre** and is inappropriate for quality fibre communications. However, a large group of multimode fibres allows light to be transmitted and received for imaging, as in optical fibre endoscopy medical applications [1.13]. As fibre diameter is reduced, a point is reached where only a single mode is present. Fibres constructed in this way are termed **monomode** or **single mode fibres**. Monomode or single mode fibres aren't affected by intermodal dispersion and can carry Standard High (SH) bandwidth digital signals up to 50 GHz. Data rate of a digital signal depends on available bandwidth so data rates greater than 100 G bps are passed through monomode fibres.

Material dispersion (*chromatic dispersion*) may be produced by variation in propagation velocity within the wavelengths of the source spectrum. This occurs because the refractive index is a function of wavelength, and any light source, no matter how narrow in spectrum, has finite spectral width. This type of dispersion is a function of cable length and line spectrum and is given by:

$$t_{md} = c_{md} \times \Delta\lambda \times D \qquad\qquad \textbf{(eq 1.4)}$$

where: t_{md} is the material dispersion in nanoseconds (ns).

c_{md} is the coefficient of material dispersion in ns/nm-km

$\Delta\lambda$ is the line spectrum in nm, and D is the distance in km.

Typical material dispersion as a function of Near Infra Red (NIR) wavelength is indicated in figure 1.9.

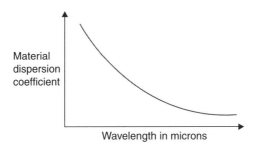

Figure 1.9: *Typical material dispersion.*

1.4.5.2 Optical fibre losses

Signals passing along optical fibres are attenuated by three key processes: imperfect reflection at the core-cladding boundary, light scattering within the fibre and absorption [1.12]. Cable attenuation is the sum of these three processes and is quoted in dB km^{-1}. Typical losses in modern cables are 0.001 - 0.7dB km^{-1}. Loss increases with damage or contamination of fibre or refracting surfaces, and is avoided by protecting fibre with cladding and another plastic protective layer or 'jacket'. Thermal agitation causes variation in core refractive index and generates scattering. The longer the wavelength, the less scattering, so NIR radiation is preferred to visible light for long-range transmission. In cold climates, fibre core/cladding refraction is affected, and more light is lost from waveguides. Scattering is caused by iron and copper glass impurities, and a very pure form of silica is used to make the fibre in order to minimise losses (figure 1.10).

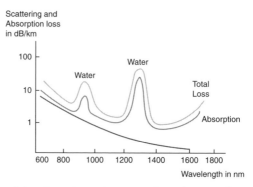

Figure 1.10: *Fibre transmission absorption and scattering dependence as a function of wavelength.*

Electromagnetic waves have strong absorption in the NIR, producing pronounced water peak absorptions. Care must always be taken to select transmission frequencies well away from these peaks if low attenuation rates are required. In addition to transmission losses, fibres suffer from jointing and fibre coupling to the transmitter and receiver in a similar way to microwave joints. Losses account for about 0.5 to 3 dB per junction and can easily outweigh fibre propagation losses, especially if there are many out-couplers to every in-coupling (see figure 1.11).

Figure 1.11: *Optical fibre splitter.*

Total loss depends on the fibre's length. Short runs around ships and aviation platforms can accept higher transmission losses as they are still small compared with jointing and coupling losses.

There are four key wavelengths used for fibre optic transmission with low optical loss levels (table 1.3):

NIR Window	Wavelength	Loss
1st Band	850 nm	3 dB/km
2nd Band	1310 nm	0.4 dB/km
3rd Band	1550 nm	0.2 dB/km
4th Band	1625 nm	0.2 dB/km

Table 1.3: *NIR transmission windows for optical fibre transmission.*

1.4.5.3 Transmitters for optical fibres

The earliest light source used with optical fibre was the Light Emitting Diode or LED, although LED cannot match the narrow bandwidth or power levels of lasers (figure 1.12). It is important to use a transmitter of sufficiently narrow bandwidth to reduce spreading to acceptable levels. For monomode fibres, low power semiconductor lasers produce an intense radiation beam with a narrow 0.1 nm spectral range, keeping spectral spreading low. Lasers are discussed in Chapter 7.

The frequency bandwidth of a narrowband optical source is given by considering the − 3 dB points in wavelength either side of the peak intensity output:

$$V_2 - V_1 = \frac{c}{\lambda_2} - \frac{c}{\lambda_1}$$ **(eq 1.5)**

Example 1.2: Determine the frequency bandwidth of an optical laser source whose FWHM is 6 nm, with − 3 dB points at 1295 and 1301 nm (2 significant figures).

$$V_2 - V_1 = \frac{c}{\lambda_2} - \frac{c}{\lambda_1}$$

According to the above equation:

$$V_2 - V_1 = \frac{3 \times 10^8}{1295 \times 10^{-9}} - \frac{3 \times 10^8}{1301 \times 10^{-9}} = 1.1 \times 10^{12} \text{ GHz}$$

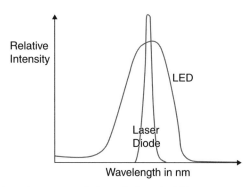

Figure 1.12: *Relative intensity outputs of a laser diode and a LED.*

A LED is assumed to radiate power as a Lambertian source, i.e. its radiation pattern has a peak at $\theta = 0$ where θ is measured with regard to the direction perpendicular to the LED surface. At other θ, the relative intensity falls off as $\cos \theta$.

For this case, coupling loss is given by:

$\eta_c = 10 \log (NA)^2$ (**eq 1.6**)

where NA is the fibre's Numerical Aperture.

> ***Example 1.3:*** Consider one such LED whose diameter is matched to a fibre core with NA = 0.15. What is the coupling loss η_c (1 decimal place)?
> In this case, coupling loss is given by: $\eta_c = 10 \log (0.15)^2 = -16.5 \, dB$

Without intermodal dispersion in monomode fibre, this allows exploitation of fibre's wide potential bandwidth. Lasers are used in Amplitude Shift Keying (ASK), where information is transmitted by shifting the radiation amplitude between fixed values (Chapter 4). Optical fibres can transmit analogue and digital signals, but digital better exploits the available bandwidth.

Optical fibre cannot match the high power-carrying capacities of waveguides, although some high power pulse applications are possible with high power lasers, e.g. carbon dioxide (10.6 microns). Bystronic UK Limited is a manufacturer of CO_2 laser cutting machines with output powers from 2.2 to 6 kW and true fibre laser machines ranging from 2 to 4 kW [1.14].

The active medium of a fibre laser is one or a combination of two rare-earth ions, e.g. praseodymium, neodymium, holmium, erbium, thulium and ytterbium ($Pr^{3+,}$ Nd^{3+}, Ho^{3+}, Er^{3+}, Tm^{3+}, Yb^{3+}) doped in silicate-based optical fibres. High power multimode diode-pumped radiation is injected through the ends of the fibre into undoped glass cladding (pump cladding), producing lasing in the rare-earth ion doped fibre core and delivering a single mode output. Use of various rare-earth ion doping materials allows operation at any wavelength from the visible to the mid-infrared – for instance, lasing in Nd^{3+}, Yb^{3+} and Nd^{3+}/Yb^{3+} doped fibres is around 1μm.

The biggest factors in terms of electrical noise in optical receivers are *shot noise* and *dark current*. Shot noise results from random arrival times of photons and the resulting discrete nature of electrical charges generated. Current generated by a photodiode is a shot noise process with intensity proportional to incident optical power. In the absence of incident light, a form of shot noise, the dark current, is generated in the detector due to thermal activity of electrical charges. With no incident photons present, I_s shot noise current is given by: $I_s = \sqrt{2q_e I_D W}$, where I_D is the dark current and W the bandwidth of the detector. With incident light, the photocurrent I_p is much larger than the dark current. Thermal noise is present in the receiver amplifier and considered 'white noise' at frequencies in the microwave band and below.

1.4.5.4 Advantages of optical fibres

Optical fibres provide several benefits:

1 Wide available bandwidth.

2 Very low transmission losses.

3 Immunity to unwanted emissions of electromagnetic waves due to electrical isolation of transmitter and receiver.

4 Fibre isolation with the jacket gives immunity to Electromagnetic Interference (EMI), especially Electro Magnetic Pulse (EMP), and freedom from leakage and crosstalk, providing security against unauthorised interception.

5 Significant savings in weight and volume compared with conventional coaxial cable and other waveguide types.

Against these advantages must be considered the disadvantage of coupling transmitters and receivers to the fibre and splicing or jointing. On balance, optical fibres provide reliable and efficient bearers for high data rate channels, and use in merchant vessels, warships and aircraft for data highways and internal communications will increase.

1.4.5 Optical Repeaters

Conventional repeaters use optoelectronic circuits to convert optical signals back into electrical signals for regeneration and amplification prior to conversion once more back into optical signals for the next stage of the signal's journey. An electronic repeater is constrained to operate at a fixed bit rate and is better replaced if the fibre optical system is to be upgraded to higher speed and avoid electronic noise being added to regenerated signals.

As optical amplifiers are based on use of semiconductor lasers or doped optical fibre, amplifiers based on *stimulated emission* are used in the laser diode. In the manufacture of laser amplifiers, cavity resonance results in a laser that acts as an amplifier. Lasing action is discussed in Chapter 6.

The future of fibre communications may be in the use of non-dispersive pulses or solitons transmitted thousands of kilometres through fibres whose loss has been compensated by optical gain, providing that all-optical systems are able to handle 100 Gbits/s [1.15].

1.5 Multi-channelling

Every communications channel can support or carry at least one independent set of information. The capacity of some bearers, e.g. optical fibre, is much greater than the capacity required to carry one set of information, whether audio or video. Such bearers can support multiple channels and the process of transmitting several information sets over the same bearer is known as multi-channelling. The total number of channels in a given system is given by the number of independent sets of information the system can support. A block diagram of a multi-channelled simplex circuit is shown in figure 1.13.

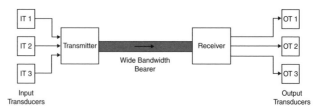

Figure 1.13: *Block diagram of a multi-channelled simplex circuit.*

> **Example 1.4:** If bandwidth of a high quality High Definition TV (HDTV) signal is 8 MHz and available bandwidth of the channel is 750 MHz, what is the *maximum* theoretical possible number of similar TV signals the channel could support?
> Maximum possible number of similar bandwidth channels = 750/ 8 = 94 channels.

Typical broadband optical cable has about 750 MHz of available bandwidth, which allows 94 possible frequencies to handle up to 38 Mbits per second of digital data (as found on a satellite transponder for inter-operability). In practice, around 40 of these frequencies are used for analogue signals, some for radio and data services and the remainder for digital TV. Digital technology allows many TV channels to occupy the same frequency space that would otherwise be occupied by a single analogue cable TV channel. Many cable providers fit about ten digital Standard Definition (SD) channels or two digital HD channels on a single analogue channel frequency using Time Division Multiple Access (TDMA), with 50 ms allocated for TV station 1 and then 50 ms allocated to TV station 2 before returning to TV station 1 and the whole process repeating. Some providers can fit even more channels on to a single frequency using various compression techniques, but this can cause degradation of channel video quality.

In principle, a channel can be divided into time slots or a number of frequency allocations. This approach gives rise to multiplexing.

1.6 Multiplexing

In multiplexing, the bandwidth of a wide bandwidth channel is controlled to allocate time slots or frequency allocations to different signal basebands. If the channel is divided into frequency allocations, the process is called Frequency Division Multiplexing or Frequency Division Multiple Access (FDMA) (figure 1.14 a); if it is divided into time allocations, it is called Time Division Multiple Access (TDMA) (figure 1.14 b).

In FDMA, the channel is divided into frequency allocations, which need not be equal in bandwidth. Placement of signals is performed with a 'mixer' and several subcarriers. FDM is suitable for transmission of analogue and digital signals. An increase in transmission rate is achieved by increasing frequency allocation, adding further bandwidth.

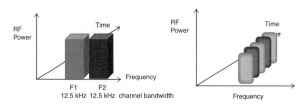

Figure 1.14a: *FDMA* 1.14b: *TDMA.*

In TDM, the channel is divided into equal duration time slots. Each signal is placed in its time slot on a bit-by-bit basis or, more commonly for PCM signals, a word-by-word basis. A given number of time slots is called a frame and if there is a need to increase transmission rate, more time slots per frame are obtained. Owing to the discontinuous nature of the time slots, TDM is only suitable for digital signals. For security reasons, modern military communications systems are encrypted to provide secure digital signals. NATO's Link 16 uses such TDMA to control short time slots on the order of milliseconds.

1.7 Basebands

A baseband is the electrical form of the original analogue information leaving the input transducer. Such a baseband may undergo considerable processing signal prior to transmission. Electrical base bandwidth may be different to the original signal bandwidth and reflect the response and quality of a particular transducer. There are many problems associated with direct transmission of basebands and it is usual to translate them to higher frequencies before transmission. The process of frequency translation is *modulation* and the baseband is called the *modulating signal*. Every signal has its own baseband, and these are divided into two distinct classes: analogue and digital.

1.7.1 Analogue basebands

An analogue baseband is a continuously varying signal of arbitrary shape or form, taking values between set limits. The simplest analogue baseband is a sine wave or sinusoid (figure 1.15).

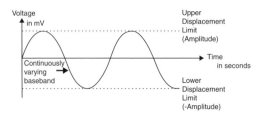

Figure 1.15: *Analogue sine wave signal.*

Real analogue basebands contain different frequencies, usually without any harmonic relationship between them, varying continuously. The signal bandwidth is given by the difference between the highest baseband frequency (f_H) and the lowest (f_L), expressed in equation 1.7 as:

Base bandwidth $= f_H - f_L$ **(eq 1.7)**

The limited response range of many input transducers to incoming signal limits practically the electrical base bandwidth output to *less* than the original signal bandwidth. Such limiting of signals to a smaller frequency range is known as band-limiting.

Example 1.5: A typical male voice signal (50 Hz to 16 kHz) in electronic communications is limited by the microphone input transducer to a relatively narrow frequency band between 300 Hz to 3.4 kHz. What is the base bandwidth of such a band-limited, intelligible voice signal?

Base bandwidth $= f_H - f_L$

$= 3400 - 300 = 3100$ Hz

1.7.2 Digital basebands

Digital basebands are discontinuous and contain virtually instantaneous transitions between clearly defined preset states. There is no limit to the number of possible states used; the most common digital baseband is a *binary* signal with two permitted states, and it is this type of digital signal discussed here. The simplest digital baseband form is a square wave with the preset states represented as previously by the wave amplitude.

The definition of a binary signal says nothing about the nature of the two separate states, but only their number. Any system that presents information by alternating a signal between two states is thus a binary system. In digital communications, it is common to consider a binary baseband as a voltage switching between two distinct voltage levels at fixed time intervals. This introduces the concept of the binary digit or *bit*, the fundamental digital signalling unit where both states have the same duration. Digital system speed is given in bits per second (bps). Modern MP3 audio format using lossy data compression has its highest level of bit transmission at 320 kbit/s, while Blu-ray Disc has a maximum data rate of 40 Mbit/s. Current terrestrial optical fibre broadband suppliers in the UK (Ofcom monitored) now provide broadband speeds for BT and Virgin respectively of 60.6–63.5 Mbits/s and 136.9–146.9 Mbits/s.

Data transmission *volume* has accelerated off the scale over the last hundred years. For much of recorded history, the ability to transmit large volumes of information was hindered by the inability to switch a transmitter on and off fast enough to transmit data. Flashing shields in Greek warfare or smoke signals by Native American Indians limited data transmission mechanically to a few bits per minute. Even modern-day use of an Aldis lamp at sea only increases this by an order of magnitude at best.

The latest optical fibre switched experimental telecommunications systems now provide extremely reliable switching rates.

A typical binary baseband is shown in figure 1.16.

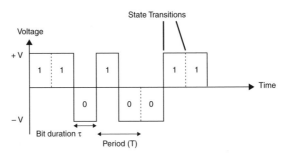

Figure 1.16: *Typical digital baseband.*

Note: Vertical lines drawn between 1 and 0 transitions are included, although they don't exist in practice. Dashed lines denote adjacent 1 to 1 and 0 to 1 bit transitions respectively.

> **Example 1.6:** If the bit duration $\tau = 2$ ms, what is the period?
> Period $= 2\tau = 2 \times 2$ ms $= 4$ ms.

1.8 Noise, amplifiers, filters and signal to noise (S/N) ratio

Until now, we have considered a perfect or ideal channel. In reality, all signals are subject to transmission loss, or attenuation, with noise added as signals pass through the channel, especially through the bearer, although noise can be introduced in the transmitter and the receiver itself. Noise is considered a natural background against which signals are read. Noise sources are grouped into two main headings: *external noise* (originating in the bearer) and *internal noise* (originating in the receiver).

1.8.1 External noise

External noise is generated outside the receiver and may be classed into three groups: atmospheric, man-made and galactic noise.

1) Atmospheric noise, or *static*, is a randomly varying signal with short, unpredictable, large amplitude bursts superimposed on it. The main source of atmospheric noise is from lightning and can be severe during terrestrial thunderstorms; it is usually observed below 20 MHz frequencies.

2) Man-made noise is generated by electric devices such as motors or power transmission lines, etc., and is often a problem in urban and industrial areas. For this reason, sensitive radio receivers are usually placed in rural areas away from main sources of man-made noise. International regulations limit man-made radio noise by use of suppression devices on the most common or severe sources. The intensity of man-made noise decreases with frequency and is mainly found between 20 and 40 MHz, although these values depend upon how near noise sources are.

3) Galactic noise is radio noise originating outside Earth's own atmosphere, i.e. unwanted 'extraterrestrial' electromagnetic radiation. Such noise decreases with increasing frequency, and is the main noise source between 40 and 250 MHz.

1.8.2 Internal noise

Above 250 MHz, noise generated in the receiver is the strongest noise source and generated by changes in performance of mainly active circuit elements, or random thermal changes in passive circuit elements. Thermal noise is ultimately the most important noise source and sets the limits on receiver sensitivity. As the name suggests, thermal noise is generated by the temperature variation of passive circuit elements, which can be improved by cooling. All receivers operate against this noise background and elimination of noise is a key receiver requirement. Details about noise sources can be found elsewhere [1.12].

1.8.3 Amplifiers

As signals propagate or travel through a bearer medium (i.e. the atmosphere), they suffer transmission loss. Attenuation is caused by several factors – geometrical spreading, absorption and scattering – so each transmitted signal steadily reduces as it travels further away from the transmitter.

As receivers require a minimum signal input, it is important where possible to increase the signal size. Modern electronic systems use a variety of amplifiers to achieve this signal increase. Amplifiers are devices that increase the signal size. The circuit symbol for an amplifier is shown in figure 1.17.

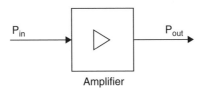

Amplifier

Figure 1.17: *Block diagram of an amplifier showing input and output power levels.*

Amplifier power gain is simply defined as the output power ratio compared with the input power ratio, i.e. amplification gain $= P_{out}/P_{in}$, but is usually quoted in decibels (dBs) in engineering terms.

Example 1.7: The input power to a loudspeaker is 0.5 mW. The output power of the loudspeaker is 100 W. Find the amplifier gain and the dB gain of the loudspeaker amplifier used (2 significant figures).

$$Gain = \frac{P_{out}}{P_{in}} = \frac{100}{0.5 \times 10^{-3}} = 200000$$

$$Power\ gain\ in\ dB = 10\ log\left(\frac{P_{out}}{P_{in}}\right) = 10 \times log\left(\frac{100}{0.5 \times 10^{-3}}\right) = 53\ dB$$

Since an amplifier amplifies both the wanted signal *and* any unwanted noise, it is important that care should be taken to limit or reduce noise inputs to amplifiers. High quality amplifiers should be used if possible.

1.8.4 Filters

Real signals contain a range of frequencies known as the signal bandwidth. Noise, however, occurs across a much wider frequency spectrum. To restrict the frequencies passing through a circuit, and restrict the noise, a frequency limiter or *filter* is used. A filter is a device whose gain (amplification) varies with the frequency of the signal passing through. An ideal or perfect filter has a perfectly flat top and vertical sides, but in practice some rounding of shape occurs due to a circuit always taking time to respond to the incident signal (inertia). A typical frequency response of a Band Pass Filter (BPF) is shown in figure 1.18 alongside the corresponding ideal response for comparison.

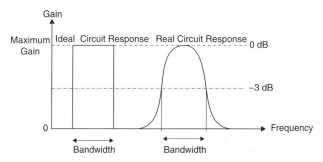

Figure 1.18: *Response of both an ideal circuit and a real circuit.*

Filter bandwidth is taken to be the frequency range for which gain is within 3 dB of the *maximum* gain (–3dB is equivalent to a factor of one half, or half the original signal level), and is a relative measurement between two similar things. This is the 3 dB bandwidth, and is the most common filter bandwidth used. A BPF only passes a band of frequencies. Filters are constructed, however, that will pass all frequencies above a cut-off frequency High Pass Filter (HPF), all frequencies below a cut-off frequency Low Pass Filter (LPF), or all frequencies outside a specified band, Band Stop Filter (BSF). The frequency responses and circuit symbols of these four common filter types are shown in figure 1.19.

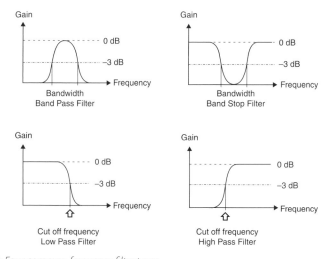

Figure 1.19: *Four common frequency filter types.*

Filters can be made using capacitors, inductors, resonant cavities, physical materials (crystals) or with computer programs (digital filters). A tuneable filter is one where the various pass or cut-off bands can be varied in frequency. A BP filter (either fixed tuned or tuneable) is often used as the tuned circuit for tuned amplifiers.

1.8.5 Signal to noise (S/N) ratio

All signals are read against a background of noise, whether radar, radio, light or acoustic sonar. If we consider a signal that spreads out from a transmitter, the electromagnetic wave intensity will vary with the inverse square of distance. Noise is generally fairly constant regardless of where you are and appears as a randomly varying intensity, as shown in figure 1.20.

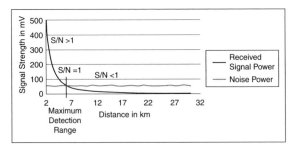

Figure 1.20: *Inverse square law and signal to noise ratio.*

A signal can be received and read correctly provided the receiver signal power is greater than the receiver noise power, i.e. the signal to noise ratio (S/N) is greater than 1 (S/N > 1). This is found by dividing the signal power level by the noise power level. Some modern transmission systems, such as Direct Sequence Spread Spectrum (DSSS), can recover even after it has fallen below the ambient noise level.

For normal signal transmissions, if the S/N < 1 amplification doesn't improve the S/N ratio as noise is amplified with the signal and the ratio between both remains unchanged. In practice, amplifiers introduce noise of their own and the S/N will be reduced further!

Example 1.8: A low frequency transmitted signal falls off with distance as $S = 10 - \dfrac{4}{R} - \dfrac{6}{R^2} \mu V$ under normal weather conditions due to free space spreading (inverse square distance) and surface interactions (inverse distance), where R is the distance in kilometres. The noise level is at a constant background of 2µV.

Calculate at what range the S/N = 4.

Since $S = 10 - \dfrac{4}{R} - \dfrac{6}{R^2}$:

$\dfrac{S}{N} = \dfrac{10 - \frac{4}{R} - \frac{6}{R^2}}{2} = 4$

$\dfrac{S}{N} = 5 - \dfrac{2}{R} - \dfrac{3}{R^2} = 4$

$5 - \dfrac{2}{R} - \dfrac{3}{R^2} = 4$

$(5 - 4) - \dfrac{2}{R} - \dfrac{3}{R^2} = 0$

$1 - \dfrac{2}{R} - \dfrac{3}{R^2} = 0$

$R^2 - 2R - 3 = 0$

Factorising: $(R + 1) \times (R - 3) = 0$

Taking the positive solution of distance for a real solution:

$R - 3 = 0$

Thus $R = 3$ km.

1.9 Networks

Networks consist of four key elements: bearers, switches, repeaters and user stations. A user station may be a laptop or Wi-Fi phone. User stations are not discussed in this book.

A switch is any system that directs signal traffic analogous to a telephone network exchange. Modern switches (or servers) are powerful computers automatically routing traffic to intended destinations. Some switches are more important than others and some are also used to connect or interface with networks, usually designated as *nodes*.

A switch is described by its message-handling capability, which is in turn determined by its transmission speed, processing speed and storage capacity. In general, the faster the transmission and processing speeds, the lower the requirement for traffic storage at a switch.

If the distance between switches, or between a user and a switch, becomes too large, a signal may be severely attenuated and become too weak for efficient communication. To overcome this, *repeaters* are inserted into the bearer; they are effectively amplifiers that increase the signal size after it has been attenuated. On ships few are used, although larger platforms such as large passenger vessels have greater requirements. In the past, only vessels with large numbers of passengers could justify the business case to provide personal on-board and roaming mobile communications, but today passengers and crew expect access to personal mobile communications anywhere and at any time. A new generation of satellites is being developed to provide global personal communications, including Iridium, Globalstar and Orbcomm, the latter requiring several orbital planes and a combination of Low Earth Orbit (LEO) and Middle Earth Orbit (MEO) satellites. Global Mobile Personal Communications by Satellite (GMPCS) are personal communication systems providing transnational, regional or global coverage from a constellation of satellites accessible with small and easily transportable terminals. As yet, limited provision is available. GMPCS services include two-way voice, fax, messaging, data and broadband multimedia [1.16]. The author proposes that in future, with server satellites in LEO and MEO orbits, suitably protected or 'hardened' against solar flares and Sudden Ionospheric Disturbance (SID), there will be new Space Based Web (SBW) services, also providing infrastructure for future lunar and extraterrestrial exploration.

Repeaters are found on radio links, copper cables and all optical amplifiers. As the signal is attenuated the S/N ratio falls and a digital optical repeater restores S/N to an acceptable level. Simple non-processing electronic repeater amplifiers cannot do this as they amplify noise and signal. The reason all optical repeaters can do this is because it is a *processing repeater*, which reduces the signal to its baseband, 'cleans it up' and then retransmits it in its original form.

There are a number of different simple networks: bus, star, ring, meshed, fully connected networks and tree networks. Of these, we will look at the most relevant.

1.9.1 Star networks

A star network consists of one central switch, *hub* or computer, which enables the transmission of messages between different nodes. The central node is connected to all other nodes and provides a common connection point for all nodes through the hub. Failure of the central hub node means all other connected nodes cannot communicate with each other. Failure of a link between any of the users and the switch results in isolation of that user from all the others, but one advantage is that the rest of the system will be unaffected [1.12], while the fact that no user can interact with another without going through the switch or server means that security is high. This single star network switch (figure 1.21) links a number of users and routes signal traffic between the various users. As these switches allow users access to the network, they are known as 'access' switches and form the lowest level of switch in a network (e.g. a local exchange in the BT network).

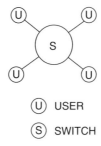

U USER

S SWITCH

Figure 1.21: *Star network.*

As the number of users increases, or the geographical spread of users becomes greater, further switches are introduced. As the number of switches rises significantly we need to consider two significant forms of network: linear and meshed (gridded) networks.

1.9.2 Linear or serial networks

A linear network (figure 1.22) is a collection of switches where each switch is connected to two other switches with no branching or interconnections, linked in a 'daisy chain'. Signal traffic flows linearly along the network and each switch is an access switch for local users.

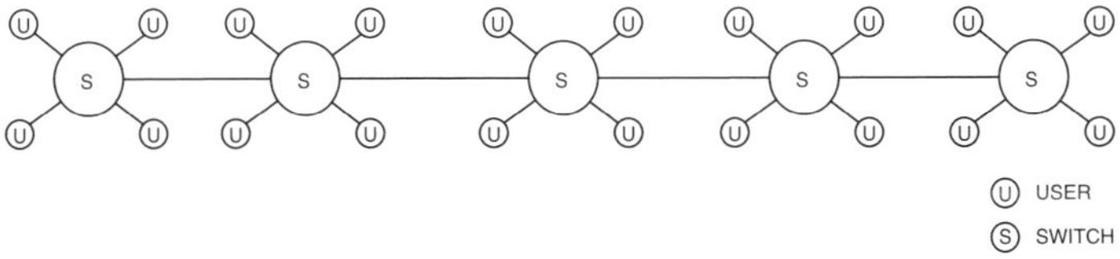

U USER
S SWITCH

Figure 1.22: *Linear network.*

A major drawback is that if a switch or connection is broken, some or even all the network becomes inoperable. For example, if the middle switch in figure 1.23 stops working, the network will be broken into two separate fragments, both of which may be unable to act as local, independent networks.

1.9.3 Meshed or gridded networks

The usual solution to this problem in large networks is to build in bearer *redundancy*, providing interconnections between switches. This type of network is a 'meshed' or 'gridded' network and provides increased survivability to the network, introducing *graceful degradation* instead of catastrophic failure (figure 1.23).

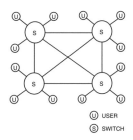

Figure 1.23: *Meshed or gridded network.*

Redundancy provides alternative links so no single broken inter-switch connection prevents traffic flow. This arrangement prevents overloading of a single network switch, sending some traffic on alternative routes. At a national level it is impractical to link all the access switches, so a 'hierarchical' network is used instead, having a central core (or inner ring) of meshed nodes, with an outer ring of access switches connected to these nodes. It is impractical to fully mesh all these nodes, so instead the nodes should be connected to at least two other nodes as you would find in a telephone network, comprising local exchanges (access switches), meshed regional exchanges (nodes), and a few international exchanges, gateway nodes to other networks. Users wanting to make local calls only use their access switch and don't need to place traffic on the central meshed core. Further discussions about terrestrial wireless networks are found elsewhere [1.17].

1.10 Switching

The method by which nodes and switches establish communications between users is the switching method. Switching is largely controlled by whether a communication link must be real-time (continuous link between two users, e.g. telephone) or non-real-time (discontinuous path between two users, e.g. computers). The main real-time switching technique is *line switching* and requires an unbroken line through the network established between both users. The link must remain connected as long as the communication between those two users is needed. Line switching doesn't require message storage capacity at the switches, although most switches provide some storage capacity, but it does require an entire line to be established before communication starts and this is performed by prefacing the call with the unique dialling code of the user to be called.

The main form of non-real-time switching is *message switching*, where each signal starts with the routing information as an address. The message is sent to the user's access switch, which then opens a line to the next switch or node en route. The signal finds its way from switch to switch until it arrives at the intended recipient. A router is an internet working device that forwards packets between networks by processing the routing information.

Message switching differs from line switching in that the route from sender to recipient is opened one section at a time. Messages find their way switch to switch through the network in a similar way to a letter moving from sorting office to sorting office in a mail system. Since complete lines are not opened, a message switching system has a high message handling capacity, although it doesn't have a real-time capability.

A modern variant of message switching is *packet switching*, where a message is broken into equal sized packets and these are independently routed through the network. Arriving at the recipient's access switch, the packets are reassembled into the message and sent out to the intended user. Packet switching improves a network's performance, allowing nodes to equalise their bandwidth loadings on the various switched routes. When one user is not sending packets, the link can be filled with packets from other users.

1.11 Computer networks

A computer network is a network allowing computers to exchange data with one another; the best-known computer network is the Internet. In computer networks, networked devices exchange data with each other along network links. Connections between nodes are established using either cable media or wireless media.

1.11.1 The Internet

The internet is a global system of interconnected computer networks that uses the Transmission Control Protocol Internet Protocol (TCP/IP) suite to link billions of devices around the world [1.18]. It is a network of networks, consisting of millions of private, public, university, commercial and government networks from local to global scope, linked by a wide range of electronic, wireless and optical networking solutions. The internet carries an extensive range of information resources and services, including mobile applications, social media, web browsing of hypertext documents and applications of the World Wide Web, electronic mail, online gaming, telephone calls, video conferencing applications such as Skype, and peer-to-peer document sharing networks. Tim Berners-Lee (now Sir Timothy), an English computer scientist, made a proposal for an information management system in March 1989, and implemented the first successful communication between a Hypertext Transfer Protocol (HTTP) client and server via the internet around mid-November of 1989 [1.19–1.21].

The origins of the internet date back to ARPANET, founded by the United States government in the 1960s with the intent to have a globally distributed network system that would still operate after a major global conflict. The internet has been widely used by academic communities since the 1980s. As of 2015, 38 per cent of the world's human population has used the internet within the past year, 100 times more people than were using it in 1995 [1.22].

Each network works to a particular 'protocol', which specifies parameters such as switching method, packet size, transmission speeds and the appropriate digital keying method (Chapter 4). To access a network, all equipment must follow the same protocol. Different networks, however, often use different protocols and this can lead to problems in connecting the networks. Problems are resolved by using a special node as a gateway between the two networks. This gateway node has the specialist hardware and software capabilities to 'translate' from one protocol to the other and allows signals to pass between the different networks.

Packet-switching networks such as ARPANET and other networks were developed in the late 1960s and early 1970s using various protocols. The first two nodes of what would later become ARPANET were connected between the University of California and the NLS system at SRI International in Menlo Park, California on 29 October 1969 with an initial 50 kbit/s circuits [1.23].

It is likely that, in future, new ways to provide wireless communications services will emerge. Some engineers have suggested that use of airships or aircraft in the stratosphere at high altitudes of 17–22 km over a city or port could provide a technology known as High Altitude Platform (HAP) or Stratospheric Platforms (SP) but as yet no real engineering solutions exist to provide ever wider wireless bandwidth requirements [1.24]. A proposed HAP could provide typical station coverage of up to 200 km, as opposed to a terrestrial wireless network with a station coverage of only a few kilometres. Hybrid systems with HAP augmentation is a more likely future scenario, with systems based on optical-fibre and terrestrial wireless networks merged with satellites and high altitude platforms to provide high data rate global mobile communications [1.25].

Self-assessment questions

1.1   Outline the *reasons* for communication and the way in which it has developed in the maritime environment.

1.2   Draw a diagram of a typical communications channel and describe the *purpose* of each block.

1.3   Explain the difference between simplex, half-duplex and full duplex communications channels, giving one example of each. Explain with the aid of a diagram what is meant by multi-channelling.

1.4   Explain what is meant by a baseband and explain the difference between analogue and digital basebands. For a baseband voice signal (1 kHz to 15 kHz) what is the base bandwidth of such a band-limited, intelligible voice signal? (2 significant figures).

1.5   Explain the common sources of noise in receivers, differentiating between internal and external noise and the frequency range over which each source dominates.

1.6   A low frequency transmitted signal falls off with distance as

$S= 12 + \dfrac{1}{R} - \dfrac{3}{R^2}\mu V$ under normal weather conditions due to free space spreading (inverse square distance) and surface interactions (inverse distance), where R is the distance in kilometres. The noise level is at a constant background of 2μV.

Calculate at what range the S/N = 6. What possible solution do you suggest to achieve increased range?

1.7   Draw the symbol of an electronic amplifier and state its gain in dB.

An amplifier has a power gain of 35 dB and the signal power input is 1 mW. What is the signal power output from the amplifier (2 decimal places)?

1.8   Draw meshed, star and linear networks. Explain their various advantages and disadvantages.

1.9   State the correct bearer choice for the following frequency ranges: Near Infra Red (NIR) light, microwave, voice 3 kHz signal and 160 MHz.

1.10   State the advantages and disadvantages of the different bearer types chosen in question 1.9.

REFERENCES

[1.1] 'Acoustic Telephones', Bill Jacobs, TelefoonMuseum.com website (retrieved 15 January 2013).

[1.2] 'Captain Henry Jackson's Radio Experiments', Saltash & District Amateur Radio Club (SADARC), www.g0akh.f2s.com/SADARC/index.php (retrieved 1 December 2012).

[1.3] The British Admirals of the Fleet 1734–1995, Tony Heathcote (Pen & Sword Ltd, 2002, ISBN 0–85052–835–6), p. 126.

[1.4] www.ukho.gov.uk/ProductsandServices/PaperPublications/Pages/NauticalPubs.aspx

[1.5] Microwave and Optical Waveguides, NJ Cronin (CRC Press, 1995, ISBN 978–0–7503–0216–X).

[1.6] Roman Glass: reflections on cultural change, SJ Fleming (University of Pennsylvania Museum of Archaeology and Anthropology, Philadelphia, 1999, ISBN 978–0–9241–7172–7).

[1.7] 'Optical Maser Action of Nd+3 in a Barium Crown Glass', E. Snitzer, Physical Review Letters, Volume 7, Number 12 (December 15, 1961), pp. 444–446.

[1.8] 'Dielectric-fibre surface waveguides for optical frequencies', KC Kao and GA Hockham, Proc. IEE 113 (7) (1966), pp. 1151–1158.

[1.9] 'A low threshold tunable CW and Q-switched fibre laser operation at 1.55 µm', RJ Mears, L Reekie, SB Poole and DN Payne, Electron. Lett. (1986), 22, pp.159–160.

[1.10] 'Lightwave Communications: The Fifth Generation', Emmanuel Desurvire, *Scientific American* (January 1992), pp. 96–103.

[1.11] 'Photonic crystal fibers', P St J Russell, (review article) *Science* 299 (2003), pp. 358–362.

[1.12] *Digital Transmission Systems* 3rd edition, David R Smith (Springer Science and Business Media, 2003, ISBN 978–1-4020–7587–2).

[1.13] 'History of the instruments for gastrointestinal endoscopy', James M Edmonson, *Gastrointestinal Endoscopy*, Volume 37, Supplement 2 (1991), pp. S27–S56.

[1.14] www.industrial-lasers.com/articles/2013/11/fiber-versus-co2-laser-cutting.html.

[1.15] 'Demonstration of soliton transmission over more than 4000 km in fibre with loss periodically compensated by Raman gain', LF Mollenaeuer and K Smith, *Optics Letters*, Vol. 13, Issue 8 (1988), pp. 675–677.

[1.16] 'New Satellites for Personal Communications', John V Evans, *Scientific American* (April 1998), pp. 60–67.

[1.17] 'Terrestrial Wireless Networks', Alex Hills, *Scientific American* (April 1998), pp. 74–79.

[1.18] en.wikipedia.org/wiki/Transmission_Control_Protocol.

[1.19] *Weaving the Web: The Original Design and Ultimate Destiny of the World Wide Web by Its inventor*, Tim Berners-Lee with Mark Fischetti (Orion Business, 1999, ISBN 0–7528–2090–7).

[1.20] 'Berners-Lee's original proposal to CERN', World Wide Web Consortium, March 1989 (retrieved 25 May 2008). www.w3.org/History/1989/proposal.html.

[1.21] 'Long Live the Web', Tim Berners-Lee, *Scientific American*, 303 (6) (2010), pp. 80–85.

[1.22] 'Internet use is growing rapidly', The Open Market Internet Index Treese.org. 1995–11–11 (retrieved 15 June 2015).

[1.23] 'Internet Began 35 Years Ago at UCLA with First Message Ever Sent Between Two Computers', Chris Sutton, UCLA (archived from the original on March 8, 2008).

web.archive.org/web/20080308120314/http://www.engineer.ucla.edu/stories/2004/Internet35.htm

[1.24] 'Broadband', S Karapantazis and FN Pavlidou, IEEE *Communications Engineer* (April/May 2004).

[1.25] 'Telecommunications for the 21st Century', Joseph N Pelton, *Scientific American* (April 1998), pp. 68–73.

2

Atmospheric Propagation of Electromagnetic Waves and Antennas

'Perhaps some day in the dim future it will be possible to advance the computations faster than the weather advances and at a cost less than the saving to mankind due to the information gained. But that is a dream.'

Lewis Fry Richardson (1881–1953) Preface Weather Prediction by Numerical Process *(1922), Cambridge, The University press [2.1].*

2.1 Atmospheric propagation of waves near a surface

We will investigate atmospheric propagation of electromagnetic waves primarily in terms of reflection and diffraction propagation modes, and the 'best' aerial or antennas to use. Consider first an isotropic radiator, one that radiates equally in every direction, close to Earth's surface (figure 2.1). We observe that some of the radiated waves strike the surface. Whether reflection or diffraction occurs for any individual wave striking the surface depends on several factors: the radiating antenna height, radiation frequency and the incident radiation angle to the surface. In practice, it is observed that there are three distinct types of wave: a direct wave, a ground reflected wave and a surface (diffracted) wave.

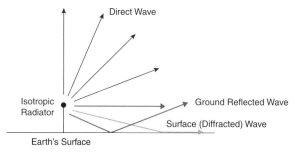

Figure 2.1: *Different wave type interactions from a reflecting surface such as the Earth.*

These three types of wave are combined to provide the three main propagation modes: ground waves, sky waves and space waves.

The constituent components of these three wave types, and importantly how they *combine*, are given in table 2.1, which includes the frequency ranges over which each mode is dominant. In the absence of an atmosphere, which can refract certain frequencies under special conditions, there will only be ground and space waves present, such as would otherwise occur on the surface of the moon.

Ground wave	Surface wave + (direct and ground reflected waves within line of sight)	Up to about 1.5 MHz
Sky wave	Direct wave + ground reflected wave, both reflected from the ionosphere	Daytime 3.0 MHz–30 MHz Night-time 0.5 MHz–1.5 MHz
Space wave	Direct wave + ground reflected wave	Generally above 30 MHz

Table 2.1: *Ground, sky and space wave frequency ranges.*

It is useful to consider the way different modes *apply* to propagation in different parts of the electromagnetic spectrum.

Table 2.2 shows the radar and radio parts of the electromagnetic spectrum, together with their main propagation modes and uses. The frequency bands are assigned by the International Telecommunications Union (ITU). However, these frequency ranges don't coincide with all maritime users [2.2].

Abbreviation	Frequency range Hz	Wavelength (in vacuum)	Modes of propagation	Uses
ULF (Ultra Low Frequency)	3–300 Hz	10^6–10^7 m	Ground wave	Submarine broadcasts
ELF (Extremely Low Frequency)	300–3000 Hz	10^5–10^6 m	Ground wave	Submarine broadcasts
VLF (Very Low Frequency)	3–30 kHz	10^4–10^5 m	Ground wave	Submarine broadcasts
LF (Low Frequency)	30–300 kHz	10^3–10^4 m	Ground wave	Radio navigation Broadcasting Time signals

MF (Medium Frequency)	300–3000 kHz	10^2–10^3 m	Ground wave Night-time sky wave	Radio navigation Broadcasting
HF (High Frequency)	3–30 MHz	10^1–10^2 m	Ground wave Sky wave	Broadcasting long-range transmissions
VHF (Very High Frequency)	30–300 MHz	1–10 m	Space wave Ionospheric scatter	TV broadcasts FM broadcasts Mobile and point-to-point links
UHF (Ultra High Frequency)	300–3000 MHz	100–1000 mm	Space wave Tropospheric refraction Tropospheric scatter	TV Broadcasts Radar Radio navigation Satellite communications
SHF (Super High Frequency)	3–30 GHz	10–100 mm	Space wave Tropospheric scatter	Radar Satellite communications
EHF (Extremely High Frequency)	30–300 GHz	1–10 mm	Space wave	Research

Table 2.2: *ITU Frequency Band allocation of frequencies and some of their typical maritime-related uses.*

2.2 Ground wave propagation

The ground wave consists of a surface (diffracted) wave, a direct wave and a ground reflected wave up to the radio horizon. At low frequencies (up to MF), effective antenna height is much less than the radiated wavelength; then the path difference in terms of wavelength between the direct and ground reflected waves becomes very small. The two waves are almost equal in magnitude but correspondingly opposite in phase (anti-phase). There is a 180-degree phase shift on reflection at low angles of incidence and combination of these waves tends to *cancel*. The ionospheric D-region (sky waves section, Chapter 2) absorbs most of the radiation striking it at these frequencies. The result at low frequencies is that only the surface diffracted wave component remains. However, as frequency increases, the direct and ground reflected waves no longer cancel exactly.

Diffraction, the change in direction of a surface wave due to its velocity when encountering an obstacle at the Earth's surface, causes waves to follow that surface.

The amount of diffraction occurring depends on frequency, and as frequency rises more waves are reflected than diffracted. The fraction of radiated power in the surface wave falls as radiated frequency rises. As the surface wave remains on the surface there is no vertical spreading, as would happen in free space, and signal intensity follows an inverse distance law rather than an inverse square distance law, as happens in free space propagation (see Chapter 1).

The wavefront at Earth's surface can induce electric currents to flow in the surface beneath it. As neither the ground nor the sea are perfect conductors, this results in a continuous drain on surface wave intensity as energy is absorbed and converted into heat. For a given radiated frequency and power, the surface wave has a greater range over sea than over desert (owing to sea being a better conductor). Nonetheless, the surface absorption creates an effective *downward* energy flow component, causing the wavefront to tilt slightly further down. As the wave proceeds, forward energy flow gradually becomes less as the tilt increases (illustrated in figure 2.2.)

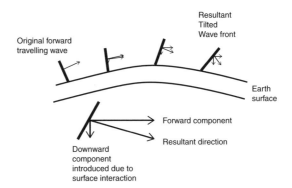

Figure 2.2: *Effect of the Earth's surface on the wavefront and propagation direction.*

Eventually so much power is lost that the wavefront finally collapses. The range achieved by a surface wave is directly proportional to transmitted power and very large power levels are required to achieve appreciable range. Range is inversely proportional to frequency and a typical surface wave transmitter must operate with high power LF signals. The surface wave dominates all radio frequencies up to about 3 MHz. There is an ambiguous transition to the sky wave region between 2 and 3 MHz, with the surface wave the dominant propagation mode for VLF to MF frequency bands (which includes GMDSS NAVTEX frequency 518 kHz and MF voice communications on 2182 kHz and Digital Selective Calling (DSC) on 2187.5 kHz).

There are two key advantages of a surface wave. Firstly, as it follows the surface of the Earth, no gaps will exist in the surface coverage so that a surface wave can form a very reliable propagation mode out to its maximum range, where the wavefront finally collapses. This component can be detected by ships if they are within the surface wave range. However, abrupt termination of a surface wave means there is a *maximum* reception range, regardless of the size or sensitivity of listening antenna. By proper selection of radiated frequency and power, Limited Range of Intercept (LRI) signals are created, very useful for tactical military communications in the MF and HF bands.

Ground waves provide a useful and reliable propagation mode for low frequencies, giving gapless cover and a definite maximum range, although large transmitters and transmitting antennas are needed.

At frequencies where wavelengths are similar to the height of the bottom of the D-region ($\lambda > 50$ km, $f < 6$ kHz), the D-region starts to behave as a near perfect conductor and waves become trapped as in a waveguide. High powers are required, but worldwide coverage is possible. This 'waveguide bounce' effect is the main communications mode at VLF and below, the Schumann resonance (www.vlf.it/Schumann/schumann.htm).

ELF and VLF provide the principal means of communicating with submerged vessels with short loop aerials attached to buoys. These long wavelength waves penetrate water better than shorter wavelength ones. The lower a wave's frequency, the less energy is absorbed by the sea. The physical basis of this phenomenon is known as the 'skin effect'. The amplitude of electromagnetic fields in a conductor decreases exponentially, thus:

$$E(d) = E(0) \times e^{\frac{-d}{\delta}} \qquad\qquad \textbf{(eq 2.1)}$$

where E(d) is the electric field at depth d, E(0) the field at the surface (i.e. $d = 0$) and δ a constant (the skin depth), which depends on the conductivity σ and wave frequency f. For waves over water:

$$\delta = 503^2 \sqrt{\sigma}\, f \qquad\qquad \textbf{(eq 2.2)}$$

> **Example 2.1:** If the skin depth has a value of 4 metres at about 10 Hz, this means the field strength at a depth of 2 metres will be $E(d) = E(0) \times e^{\frac{-2}{4}} = 0.6E(0)$ of the surface value, to 1 decimal place, a dramatic fall-off in field strength. Further details are found in references [2.3–2.4].

2.3 Sky wave propagation

Sky wave propagation is the long-range ability of the ionosphere to return a broad frequency range in the HF band during the day, and a narrower lower range in the MF band during the night. The long-range capability of HF propagation makes HF one of the most commonly used sources of time and frequency dissemination [2.5–2.6].

In figure 2.3, some direct waves are emitted up and enter the ionosphere, a region of ionised gases extending from about 50 km up to 400 km in altitude, of different electron density, due to the ionising ability of solar radiation and charged particles. But what is the effect of these layers on radio wave propagation? Radiation and charged particles striking the upper regions of Earth's outer atmosphere can ionise gas atoms. However, at low altitudes there are insufficient levels of energetic radiation (especially UV, which is absorbed by ozone at higher altitudes) but many gas molecules, while at high altitudes there are high radiation levels but few gas molecules to be ionised. So there *should* be a peak in ionisation somewhere between these extreme limits where conditions are optimum. In practice, density wave fluctuations are observed as atmospheric stratification results in molecules layering out according to molecular density under gravity, with oxygen lower than the two common nitrogen isotopes that are observed (figure 2.3).

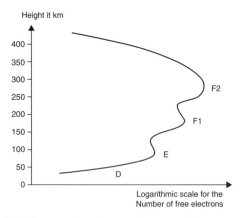

Figure 2.3: *Simple model of the ionosphere showing the main refractive regions.*

In principle, this provides three weakly layered regions of higher than normal electron density. The refractive effects of ionised gases on electromagnetic waves increases with increasing electron density and, as upward travelling waves enter a region of increasing electron density, waves are refracted *away* from the normal. As an ionised layer has its highest electron density in the centre of the layer, it follows that a wave is continuously bent away from the normal as it progresses towards the centre. This may result in waves reaching the critical angle, undergoing TIR and being reflected back towards the Earth's surface.

Should the wave *fail to* reach the critical angle by the time it reaches the centre of the layer, it passes into a region of decreasing electron density above, refracts back *towards* the normal, passes through the layer and exits above. It is unusual under normal ionospheric conditions for frequencies above 30 MHz to experience TIR and this is the upper frequency limit for sky wave propagation. As there is generally a smooth variation of electron density through the layer, the wave path is a smooth curve. Typical electron density variation with altitude is shown (figure 2.4).

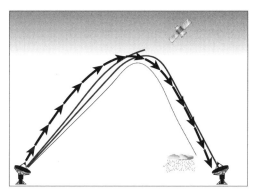

Figure 2.4: *A sky wave just returned from the maximum ionisation at the centre of the ionised layer.*

Refraction caused by a layer varies inversely with incident frequency: as frequency increases, the refraction decreases. The limit or cut-off frequency is found by transmitting electromagnetic waves vertically so they are incident on the ionosphere at normal upward incidence (figure 2.5, see plate section).

Frequencies up to a maximum value, the critical cut-off frequency, reflect back to the ground. For angles of incidence greater than zero, frequencies *higher* than the critical frequency are reflected. In December 1901, Guglielmo Marconi sent radio waves from Cornwall, England to Beacon Hill, Canada. British scientists Heaviside and Kennelly concluded that the waves in Marconi's transmission had followed Earth's curvature along electrically conductive or charged layers in the upper atmosphere. This so-called ionosphere behaved as a mirror for radio waves with wavelengths greater than about 20 metres. The British scientist Appleton studied reflections from the upper atmosphere, using continuous radio waves and the Doppler effect to detect layer movements. Several ionised layers were discovered, and Appleton suggested a division of layers ordered alphabetically, starting with the *E*-layer (Heaviside and Kennelly) at the bottom, with an *F*-layer above. Further measurements showed that the *F*-layer was divided into two parts, each with its own peak. The layers were named F_1 and F_2 respectively, the two nitrogen isotope layers. Later, a *D*-region below the *E*-layer was discovered, composed of heavier ozone O_3 molecules. The atmosphere above the *F*-layers (> 500 km) was called the magnetosphere, where the magnetic field has a major impact on movement of electrically charged particles. In 1931, Sydney Chapman presented a relatively simple mathematical model for formation of ionised layers, based on the fact that energetic photons from the sun ionise air molecules into negative electrons and positive ions [2.7].

The probability of an electromagnetic wave experiencing refraction to undergo TIR depends on the incident radiation frequency, incident initial wave angle on the layer and the layer's electron density. The result for frequencies above the critical frequency is that there is a *minimum* angle of incidence for TIR to occur. The layers of the ionosphere that cause reflection are the E, F_1 and F_2 layers. The D-region at the bottom (which disappears at night) absorbs incident radiation, and attenuation caused by this layer increases with *decreasing* frequency. For all sky wave frequencies, increasing the incident angle causes the point to be reached where the electromagnetic wave spends so much time travelling the 'thick path of the atmosphere' in the D-region (going up and coming down) that it is attenuated below ambient noise level and disappears. All sky wave frequencies have maximum and minimum angles of incidence for reflection and this creates a sky wave *reception zone* on Earth's surface (figure 2.5).

The position of the sky wave reception zone depends critically on transmitted frequency. The reflected wave at the minimum angle of incidence is the first returning sky wave and the nearest edge of the reception zone. As all other sky waves at this frequency spend longer in the D-region, they suffer greater attenuation and return weaker signals out to the last readable sky wave, marking the furthest edge of the reception zone. The distance from a transmitter to the edge of its reception zone is the *skip distance* and increases with increasing transmitted frequency. Up to this distance no sky waves are present and, beyond the ground wave, no signals are received. The region between the end of the ground wave and the first returning sky wave is the *silent zone*. Similar phenomena are found with sound waves in air and sonar underwater.

The frequency that makes the skip distance equal to the range to the receiver is the highest frequency used with a transmitter and receiver pair, which just keeps a receiver on the edge between the silent zone and sky wave reception zone. This frequency is the Maximum Useable Frequency (MUF) and is the highest frequency of least loss that may be used. In practice, the ionosphere is too unstable to allow practical use of such a precise frequency and instead we add a 15 per cent margin or 'buffer' against error by using the Frequency of Optimum Traffic (FOT), which is 85 per cent of the MUF. This frequency suffers higher loss than the MUF but is more reliable.

Example 2.2: If the MUF on a sky wave circuit is 22 MHz, what is the Frequency of Optimum Traffic (3 significant figures)?

FOT = 0.85 MUF = 0.85 × 22 MHz = 18.7 MHz. Any change in the ionosphere or range required will change the MUF.

Of these, the most important change for practical communications is *diurnal variations*. At night, the D-region completely disappears and the entire ionosphere becomes thinner and thus less ionised (due to recombination of positive and negative charges), which means the critical frequencies for the layers are reduced and consequently the MUFs are lowered. During the night, the sky wave window for a given range becomes narrower and moves down the frequency scale.

Two key effects on communications arise as a consequence of diurnal variations. The first is that, with the loss of the D-region and reduction of the MUF, some frequencies have sky wave propagation at night but are ground waves during the day. Similarly, MF stations that broadcast local signals during the day can become long-range communications at night! If these stations achieve long-range sky wave communications at night, they can cause interference in the MF band due to multipath propagation.

Additionally, if night-time signals from a wanted station arrive at a receiver by both surface and sky wave, there is a high probability that these components may not be in phase, in which case *fading* may be experienced due to changes in the sky wave path from the unstable ionosphere. For those interested in further aspects of communications fading, a reference is given [2.8].

Further key communications problems occur at dawn and dusk. At these times, ionospheric changes are *so* rapid that the MUF and the transmission frequencies change quickly as well. As sky waves give very long ranges, in some cases over 3500 km, it is vital to know where the terminator (the boundary between day and night) lies along the transmission path. In practice, it is best to avoid transmission during dawn and dusk periods of less predictable frequency.

2.3.1 Prediction of MUFs [2.9]

MUF calculation for a particular sky wave transmission is performed with tables or computer programs. In each case, the information required for the calculation is the geographical position of transmitter and receiver, and the date and time of day for transmission. This information takes into account range and latitude, diurnal, seasonal and anticipated sunspot variations of the ionosphere. Some prediction aids don't require all this information. However, these are not real-time methods and unexpected solar activity may mean expected frequencies do not work as anticipated. For further discussion of the issues of MF and HF operation 'in the field', it is instructive to read the section on High Frequency (HF) Communications in the reference following Sir Ranulph Fiennes' epic unassisted crossing of the Antarctic continent in 1992 [2.10].

Various US-developed electromagnetic-effects predictive software products, such as IREPS or AREPS, were developed for modelling propagation of electromagnetic and acoustic waves in Earth environments. However, the best prediction aids are real-time ionospheric sounders, which use a scanning receiver to read a stepped swept frequency signal sent by a transmitter. As the transmitted signal is swept in frequency over the sky wave range, the readout from a scanning receiver is a graph of all frequencies reaching the receiver and their propagation times (figure 2.6). The part of the frequency spectrum that reaches a ship at that time is the best frequency allocation to communicate with that shore station, but this will soon vary due to changes in the ionosphere. This is useful when countering the effects of Sudden Ionospheric Disturbance (SIDs), solar flares or Coronal Mass Ejections (CMEs) from the solar surface. Under these conditions, fading occurs rapidly and the effects may be so extreme that the sky wave window passes out of the HF transmitter frequency range, although sky waves may now appear at the lower end of the VHF band.

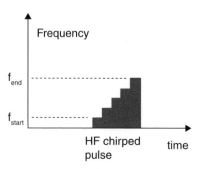

Figure 2.6: *HF transmission increasing in 3 MHz frequency steps.*

2.4 Space wave propagation

The space wave is the sum of direct and ground reflected waves. As transmitted frequency rises, antenna height can become large compared with transmitted wavelength, and the proportion of transmitted power diffracted falls. Above 30 MHz, the space wave is the main propagation mode and communication is Line Of Sight (LOS). Consider the geometric position of transmit and receive aerials (figure 2.7). The maximum range between space wave transmitters and receivers depends on the heights of both aerials. In practice, a space wave is slightly refracted towards Earth's surface, meaning the space wave range is actually 1/3 more than the geometric range, so the radio horizon is 4/3 of visual horizon. Average LOS distances are 25–30 miles, although ranges of up to 100 miles can be achieved. LOS path performance is affected by various phenomena such as free-space loss, terrain and precipitation.

Figure 2.7: *Space wave propagation travelling just beyond visible horizon.*

Depending on the transmit and receive antenna heights, different ranges are possible at sea. Using a simple online calculator [2.11], one obtains the VHF LOS range as a function of antenna height. Examples are given in table 2.1.

Transmitter height in ft	Receiver height in ft	Range in miles	Platform
20	20	10	Small vessel
200	200	32	Large vessel
2120.67 height of Warsaw Mast at Konstantynów, Poland	2120.67	102	Warsaw Radio Mast
3000	3000	120	Height easily achieved by helicopter
30000	30000	382	Satellite or high altitude aircraft

Table 2.1: *Relationship between height of transmitter, receiver and resultant range.*

Consider Table 2.1. Increasing mast height makes a significant 'footprint' or range increase. Until it collapsed in 1991, the tallest mast structure was the Warsaw Radio Mast; to go higher than this requires an aerial platform such as a helicopter or aircraft. It was no accident that after the Falklands War one of the first actions taken to prevent further sudden attacks on ships at short range was to fit Sea King helicopters with an underslung bag radar, the Searchwater radar, which when aloft above a task group provided a considerably greater detection range. Increasing height to its logical conclusion enables a space platform to have the largest footprint possible, a position satellites routinely take for telecommunications and sensing applications.

At low incident angles, interference between direct and ground reflected waves must be considered.

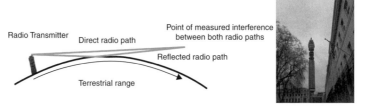

Figure 2.8: *a) Grazing incident wave interference, with tall radio transmitter tower, and b) (BT Telecom Tower, London. © CR Lavers*

With ground waves at very low incident angles, there will be a phase shift of almost 180 degrees upon reflection. The total phase difference between direct and ground reflected waves is again equal to the actual physical path difference plus 180 degrees and produces an interference pattern. As a receiver moves away from a transmitting antenna, the path difference changes and it detects a series of alternating maxima and minima as direct and ground reflected waves move in and then out of phase periodically. As range increases, path difference reduces until finally there is only destructive interference with intensity falling continuously to zero. Beyond the radio horizon only diffracted surface waves are left, which die away quickly at these relatively high frequencies.

By plotting received intensity against range for a transmitting antenna for a constant height receiving antenna, we obtain the graph as shown (figure 2.9).

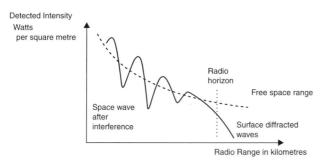

Figure 2.9: *Intensity as a function of range.*

There are also intensity variations if the receiver is moved vertically up or down. The resulting effect breaks up the theoretical uniform communications coverage into a number of 'lobes' and nulls, as shown (figure 2.10).

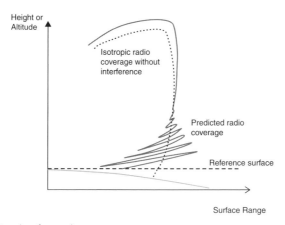

Figure 2.10: *Predicted surface radio coverage.*

In figure 2.10 there are several extended detection ranges in the lobes (constructive interference) and several reduced detection ranges (destructive interference) between lobes. Surface range is reduced as the surface represents a line of *minimum* intensity (180 degree phase change on reflection), so surface communications ranges greater than 30 km are hard to achieve in this frequency range under normal atmospheric conditions.

Even wide coverage communications systems, such as UHF systems using biconical antennas (figure 2.11), are affected by lobing. Highly directional systems, such as satellite systems, are only generally affected when looking low on the horizon so that a 'contact' and the sea surface are in the same Field Of View (FOV) at the same time. This can occur when a satellite system is on the extreme edge of its 'footprint', so a satellite in the high Arctic, for example, is just above the horizon briefly.

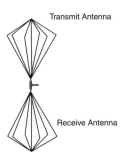

Figure 2.11: *UHF biconical antenna (transmit antenna top, receive antenna bottom).*

Lobing reduction is achieved by increasing the number of lobes present, or reducing the surface reflected component to reduce lobe formation. The number of coverage lobes is proportional to the antenna height in wavelengths, so if we increase the number of wavelengths between the antenna and surface, there are more smaller lobes with smaller gaps between them. This is achieved by raising antenna height to the top of a mast or increasing frequency to reduce wavelength. Lobe structure and surface minima are further reduced with vertically polarised antenna, as this reduces the ground reflected component.

2.5 Tropospheric and ionospheric scatter communications

Radiation can also undergo scattering from precipitation, e.g. from frequent rain or spray particles suspended in the atmosphere, or patches of high ionisation created by meteoric paths. Scattering may be back along the wave path, or in a conical region around the wave path in the forward direction. Forward scatter along the wave path can provide communications beyond the optical horizon, although scattered strength is usually low. Directional transmit and receive antennas are used, as well as a high power transmitter. Antennas are aligned so they focus on the region from which scattering is required. Scattered signal strength falls with increasing scattering angles, so it is usual to make the scattering angle as small as possible. If scattering is caused by continuous atmospheric events, scattered signal is fairly independent of atmospheric conditions, giving reliable communications, although the highly directional transmitter and receiver aerials are usually fixed.

Two frequency ranges give reliable scatter communications: 1) *UHF* Frequencies above 500 MHz are scattered by the troposphere. This tropospheric scatter gives communications up to 600 km, depending on the antennas and signal bandwidth.

2) *VHF* The second region is between 30 and 50 MHz in the VHF band. Here, the scattering medium is the E-layer of the ionosphere and ionospheric scatter can give a range of up to 2000 km; it is reliable and consistent, as meteoric bombardment goes on continuously worldwide. However, as the ionosphere is a dispersive medium, ionospheric scattering bandwidths are limited to a few kHz.

2.5.1 Tropospheric interference

In practice, the marine environment causes significant variations from *theoretical* radar Maximum Detection Ranges (MDR) obtained from the simple radar range equations discussed in Chapter 5. Extended range in some directions is due to constructive interference and reduced detection in other directions is due to destructive interference. We will now consider lobing of the vertical coverage. Many communications and sensors systems are affected.

2.6 Vertical coverage lobes

At any point P we have the typical situation shown in figure 2.12.

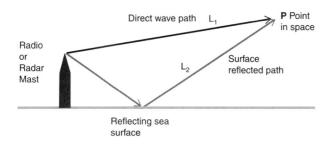

Figure 2.12: *Indirect and direct wave paths.*

Taking into account reception of waves from the two paths gives rise to an interference pattern. If the values of signal strength are plotted on a Vertical Coverage Diagram (VCD), we get lines of maximum intensity (in phase addition) and minimum intensity (out of phase subtraction). The effect breaks up theoretical uniform coverage into several lobes.

Example 2.3: What is happening in terms of path difference at the sea surface minimum?

Well, the path differerence $= L2 - L1 + \dfrac{\lambda}{2}$ **(eq 2.3)**

At sea level $L2 = L1$

Therefore the path difference $= \dfrac{\lambda}{2}$

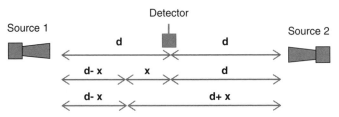

Figure 2.13: *Total phase difference. Consider two sources equidistant from a detector (d) apart.*

As the total path difference is generated by detector movement slightly towards source 1 and slightly further from source 2, the total path difference (figure 2.13) is given by: $(d + x) - (d - x) = 2x$

This results in a series of minimas (nulls) and maximas (lobes) in the communications or radar system's ability to detect targets.

The longer surface reflected wave appears to come from a 'virtual' identical source below the sea surface (figure 2.14, see plate section).

We achieve extended detection ranges in lobes and reduced detection ranges between them. Having a few large lobes has the greatest effect on radar coverage as gaps between them is large. This causes large detection range variations with target height. Surface ranges are severely reduced as the surface is a line of minimum intensity because of the 180 degree phase change on reflection. Surface ranges greater than 30 km are difficult to achieve.

Lobing generally only affects radars with large vertical beam width, such as maritime surveillance radars. Reduction of lobing effects is achieved by altering the number of lobes or reducing the surface reflected component to reduce lobe formation.

Introducing different frequencies will change the in phase, out of phase conditions, so different lobing patterns can be produced that allow surface coverage gaps to be 'filled in' (figure 2.15). Changing frequencies alters effective radar range.

Figure 2.15: *Radar lobing patterns changing as a function of frequency.*

Modern techniques fill the gaps between lobes and can include two or more vertically separated antennas, producing similar patterns but at slightly different heights, with a measure of overlap. Multiple frequency radars create different overlapping lobe patterns produced for each frequency. Radar and communications lobe pattern production is reduced with vertically polarised waveforms, as these reduce the magnitude of the ground reflected component.

In practice, in more complex situations where two adjacent sources are not radiating in phase to start with, this must be taken into account in determining the phase relationship between the waves arriving at any point P in the radiation field, and including any time delay, e.g.: Total phase difference at P = phase difference + path difference + time difference between two sources S1 and S2

$$= \text{phase difference} + \left[\frac{d_2 - d_1}{\lambda}\right] \times 360 + \left[\frac{t_2 - t_1}{T}\right] \times 360 \quad \textbf{(eq 2.4)}$$

2.7 Tropospheric refraction

Tropospheric refractive index (the bottom 15 km of the atmosphere) is a product of atmospheric pressure, absolute temperature, water vapour pressure and the ionisation. Changes of these factors with height cause significant refractive index changes. Ionisation increases slightly with height but may be taken as constant. Refractive index falls with increasing altitude, which introduces refraction *away* from the normal and bending of EM waves over the visual or geometric horizon. Under normal radar frequency atmospheric conditions, this increases range by 15 per cent more than expected from transmitter and receiver geometries alone.

However, if the air is less dense than the standard atmosphere, refraction increases and waves will strike the Earth's surface at shorter range and so a shorter detected range can be achieved (*sub-refraction*).

If air is denser than the standard atmosphere, refraction decreases and waves strike the surface at larger range and so greater detected ranges are achieved (*super-refraction*). This can occur in most places, especially the Mediterranean and Red Sea. It occurs anywhere where warm, dry air passes over cooler seas, causing a less dense atmosphere, making waves bend down. These three possible scenarios are illustrated in figure 2.16.

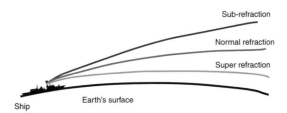

Figure 2.16: *Illustration of normal refraction (green), sub-refraction (red) and super refraction (blue).*

2.8 Atmospheric ducting

The most extreme form of super-refraction occurs when weather conditions create rapid, localised decreases in relative refractive index with height. In the case of rapid fall in relative refractive index, ducts form that can trap electromagnetic radiation to travel inside them via total internal reflection. Generally, two conditions cause this: temperature inversion, i.e. an unexpected increase in temperature with height increase, or rapid decrease in humidity with height increase.

Over the sea, humidity fall can be very rapid with a surface duct being present in most conditions except very cold, rough weather. Other weather conditions may strengthen existing surface ducts or cause elevated ducts to form. Ducting often occurs in the Arabian Gulf.

A duct acts like the ionosphere and traps radiation in it, depending on the duct size, wavelength and incident angle on the duct. As ducts prevent free space spreading, increased intensities within a duct give extended communications and radar ranges along the duct.

2.8.1 Surface ducts

The effect of a surface duct on radar forces waves to follow Earth's surface. If 1–50 cm radar is used, this gives increased surface detection ranges. A problem of surface ducts is that when a radar is mounted above the duct, radiation coupled into the duct from a ship's radar or originating in the duct from a radar stays in the duct, travelling for extended distances and not detecting anything outside the duct.

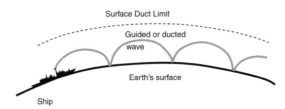

Figure 2.17: *Surface ducted waves.*

2.8.2 Elevated ducts

The main cause of elevated ducts is subsidence or sinking of a large air mass, usually associated with temperature inversion and often accompanied by rapid decrease of humidity with height. These two effects cause a rapid decrease of refractive index with height, leading to formation of elevated ducts. Elevated ducts do not give extended surface ranges, but can affect air coverage. Elevated ducts trap radiation over a range of incident angles and give reduced intensities. In the most extreme cases, gaps appear in radar coverage (figure 2.18).

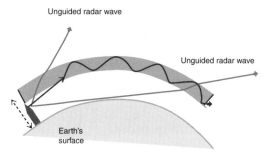

Figure 2.18: *Elevated duct.*

2.9 Communications antennas

Many reflector antenna shapes used for radars, especially dishes, are also used for communications antennas. However, there are several *whip* and wire antennas only used for communications in bands up to and including UHF. These are discussed here.

2.9.1 Dipole and monopole transmit antennas

The most common communications transmit antenna type is the so-called dipole illustrated in figure 2.19. A dipole antenna is in fact half a wavelength long and resonates at a frequency calculated from this length: $f = \dfrac{c}{2L}$ (**eq 2.5**), where L is its length.

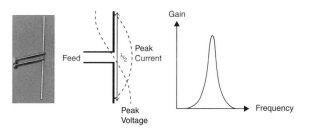

Figure 2.19: *Dipole showing current and voltage variations and gain.*

> **Example 2.4:** What is the ideal dipole length for a resonant frequency of 600 MHz (2 decimal places)? $\lambda = c/f = 0.50m$ $L = \lambda/2 = 0.25m$

To make a vessel's dipole antenna work at other frequencies it is necessary to tune it to resonate at these frequencies. Dipoles consist of two charges or elements, which receive signals of opposite polarity in each AC half cycle. A *tuned* dipole is a very efficient antenna with a narrow bandwidth. It cannot, however, be used efficiently at frequencies differing significantly from that given by its length. Dipoles are useful antennas in the VHF–UHF bands, but at HF or below are unsuitable due to their extended length. Here it is best to use a monopole antenna, mounted on the ship's superstructure, a quarter wavelength long. These are not as efficient as dipoles, but are a more convenient length. Figure 2.20 illustrates a base tuned monopole whip antenna along with its frequency response.

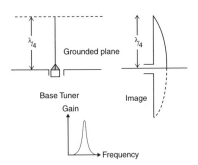

Figure 2.20: *Monopole antenna and frequency response.*

Polar vertical dipole and monopole diagrams are shown in figure 2.21.

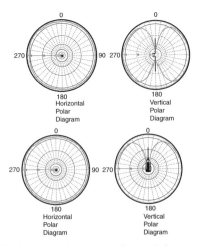

Figure 2.21: *Polar vertical dipole (top) and monopole (bottom) diagrams.*

Whip aerials are used for MF and HF. They are 7–9 metres in length and constructed of metal sections encased in glassfibre, screwed together and locked into place. Aerials are usually constructed of strong stranded copper or bronze, with ceramic or glass insulators at each end. Insulating elements prevent metal halyards from forming part of the aerial loop.

2.9.2 Highly directional Yagi-Uda antenna

This is a directional antenna consisting of multiple parallel elements in a line, made of metal rods. Yagi antennas consist of a single *driven* element connected to the transmitter or receiver with a transmission line, and further *parasitic* elements: a so-called reflector and one or more directors. It was invented in 1926 by Shintaro Uda of Tohoku Imperial University, Japan, and colleague Hidetsugu Yagi (figure 2.22).

The reflector element is slightly longer than the driven dipole, while directors are a little shorter. This design provides a significant increase in antenna directionality and gain compared with simple dipoles. The Yagi is widely used as a high-gain antenna in the HF–UHF bands. It has moderate gain and is lightweight, inexpensive and simple to construct. The bandwidth of a Yagi antenna, the frequency range over which it has high gain, is narrow, a few per cent of the centre frequency, and decreases with increasing gain. The largest and most well-known use is in rooftop fixed terrestrial television antennas and other point-to-point fixed links, and long distance shortwave communication by shortwave broadcasting stations and radio amateurs [2.12].

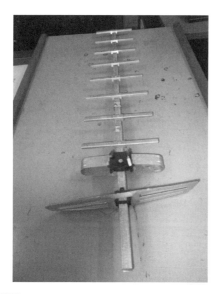

Figure 2.22: *Yagi aerial array.*

2.9.3 Broadband antennas

Dipoles and monopoles are both tuned with high efficiency and narrow bandwidth so they only send one signal at a time. Owing to the large number of signals, a maritime vessel may need a very large number of communications antennas placed on the vessel's upper decks. Consequently, a broadband wide bandwidth aerial, capable of sending several signals simultaneously, is used.

To do this, we make use of a property of tuned antenna circuits. The selectivity or quality factor (Q) helps determine the antenna bandwidth (B) from the formula:

$B = \dfrac{f_c}{Q}$ (**eq 2.6**), where f_c is the tuned centre frequency.

> **Example 2.5:** If the tuned centre frequency is 2.1 MHz and the circuit has a Q-factor of 15, what will the bandwidth B of the circuit be (2 significant figures)?
> Using the equation: $B = \dfrac{f_c}{Q}$ and substituting: $B = \dfrac{2.1 \times 10^6}{15} = 140\ kHz$

Resonant circuits have a high Q-factor and produce narrow bandwidths. The Q-factor of an antenna is a function of its height to width ratio with short, fat antennas having a low Q-factor and thus a wide bandwidth (figure 2.23).

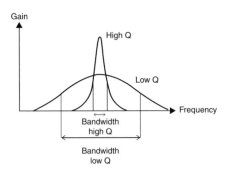

Figure 2.23: Antenna 'Q'.

In practice, both dipoles and monopoles can be made into broadband antennas by folding or 'bending' wires. An example of a multiwire folded monopole antenna based on a ship's mast is shown (figure 2.24.)

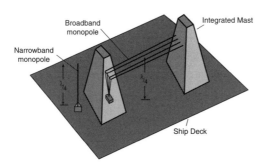

Figure 2.24: *Integrated mast with narrow and broad band monopole antennas.*

Broadband folded antennas have a lower efficiency than a resonant antenna, but don't need to be tuned over a frequency ratio of 3:1 (e.g. 2–6 MHz). Broadband antennas can send or receive several signals at the same time in the UHF.

2.9.4 Receive antennas

All communications receive antennas are broadband as they respond to any electromagnetic wave. The only requirement is that they should be long enough to gather appreciable signal strength. It is common to use receive antennas similar to the transmit antenna so we find whip HF receive antennas and dipoles and biconical UHF receive antennas.

Self-assessment questions

2.1   Discuss the various ways in which electromagnetic waves are propagated from a ship.

2.2   State the frequency ranges for ground waves, sky waves and space waves. State the common uses for the standard frequency bands.

2.3   Explain why a ground wave has a well-defined range and how it can be used for a Limited Range of Intercept transmission.

2.4   Consider a VLF wave at the Earth's surface varying according to

$E(d) = E(0) \times e^{\frac{-d}{\delta}}$ where E(d) is the electric field at depth d, E(0) is the field at the surface (i.e. d = 0). If the skin depth has a value of 10 metres at about 20 Hz, what will be the field strength at a depth of 3 metres in terms of the surface strength E(0) (2 decimal places)?

2.5   Sketch a graph of the electron density in the ionosphere against altitude, labelling all the key layers.

2.6   Draw a diagram to illustrate how the reception zone of a sky wave is formed and briefly explain it. If the critical frequency for a night-time MF sky wave frequency is 1.2 MHz on a particular sky wave link, what should the FOT frequency be (3 significant figures)?

2.7   Using diagrams, explain why there is a variation in received intensity of space waves with both changes in range and height of the receiver.

2.8   Describe dipole and monopole antennas and draw their frequency response and coverage diagrams. What is the ideal dipole length for a resonant frequency of 450 MHz (3 decimal places)?

2.9   With the aid of a diagram, explain the difference between the behaviour of high and low 'Q' resonant circuits. If the tuned centre frequency of a circuit is 7.3 MHz and the circuit has a Q-factor of 35, what is the bandwidth B of the circuit (5 significant figures)?

2.10   Explain how the bandwidth of a dipole can be increased and illustrate it with a sketch of a broadband antenna on a ship.

REFERENCES

[2.1]. openlibrary.org/books/OL7095124M/Weather_prediction_by_numerical_process

[2.2] ITU www.itu.int/en/Pages/default.aspx

[2.3] 'The ULF/ELF/VLF Electromagnetic Fields Generated in a Sea of Finite Depth by Elevated Dipole Sources', AC Fraser-Smith and DM Bubenik, Technical Report E715–2, (Stanford University 1984) www.dtic.mil/dtic/tr/fulltext/u2/a149638.pdf

[2.4] 'Sending signals to submarines', D Llanwyn Jones, New Scientist (4 July 1985), pp. 37–41.

[2.5] 'The use of National Bureau of Standards high frequency broadcasts for time and frequency calibrations', N Hironaka and C Trembath, NBS Technical Note 668 (US Government Printing Office, Washington, D.C., 1975).

[2.6] 'NBS Frequency-time Broadcast Station WWV, Fort Collins, Colorado', PP Viezbicke, NBS Technical Note 611 (US Government Printing Office, Washington, D.C., 1971).

[2.7] *Geomagnetism*, S Chapman and J Bartels (Oxford University Press, 1940, 2 volumes), 1 pp. vii–viii.

[2.8] *Digital Transmission Systems 3rd edition*, D R Smith (Springer Science and Business Media, 2003, ISBN 978–1-4020–7587–2), p. 571.

[2.9] dx.qsl.net/propagation/

[2.10] *Mind Over Matter: Epic Crossing of the Antarctic Continent*, R Fiennes (Mandarin reprinted1994, ISBN 1 856193756) p. 261.

[2.11] (www.miketooley.info/g8ckt/range.htm)

[2.12] www.packetradio.com/ant.htm

3

Analogue Modulation Techniques and Receiver Principles

'Mr Watson – come here – I want to see you.'

Written in Thomas Watson's journal (10 March, 1876). The first voice modulated line telephone call made by Alexander Graham Bell, summoning his assistant Thomas A. Watson from the adjoining room of his Boston laboratory.

3.1 Reasons for modulation

Simple basebands contain the information we want to send on our communication system. For most wireless radio and line applications it is impractical to transmit basebands directly as they are at low frequencies (i.e. a few hundreds of Hz to a few kHz). All directly transmitted basebands would propagate around the world in the duct between ground and the bottom of the ionosphere, covering large areas accessible to all with appropriate receivers.

In addition, direct baseband transmitting antennas *should* be a quarter of a wavelength long for efficient launching of waves. For a 1 kHz wave, this would mean an antenna at least 75 km long! Furthermore, the ratio of highest to lowest frequency of the signal would be very large and it would be difficult to make an antenna have the same response at such different frequencies, resulting in severe signal distortion. Although frequencies of a few tens of kHz are used for submarine broadcasts, keeping antennas to a reasonable size makes them much less than $\lambda/4$ with efficiencies under 1 per cent.

All these problems may be overcome by electrical engineering solutions, but there remains one difficulty that cannot be overcome. Most basebands occupy the same frequency range and similar basebands (e.g. voice) occupy identical range. If several voice basebands transmitted simultaneously, a receiver would be incapable of separating them, making reception an unwanted power battle, with the most powerful transmitter 'capturing' all the receivers!

To prevent these problems we must separate the basebands, typically raising them in frequency. We perform this frequency translation by combining each baseband with a wave at the required higher frequency. This wave is the *carrier wave* and the process of combining the baseband and the carrier is called *modulation*. A baseband modulates the carrier frequency. As a rule of thumb, carrier frequency should be at least ten times the signal bandwidth, e.g. voice signals of 3 kHz bandwidth cannot be sent with radio carrier frequencies below 30 kHz.

3.2 Types of modulation

To 'attach' the baseband to the carrier, we must change a fundamental parameter of the carrier wave. For modulation purposes, a parameter must be an independent variable and unaffected by changes to other variables. The basic sine wave equation consists of two independent components: amplitude A and an angle term $\sin(2\pi ft + \varphi)$, whose basic equation is shown:

$$V = A\sin(2\pi ft + \varphi) \qquad\qquad \textbf{(eq 3.1)}$$

V is the baseband voltage, f the carrier frequency, t time, and ϕ the phase.

The two independent components used for modulation are Amplitude Modulation (AM) and Angle Modulation. Angle modulation can be divided into Frequency Modulation (FM) and Phase Modulation (PM). In reality, frequency and phase are *not* fully independent and only one modulation method is used at once. If we want to increase the amount of information carried by the wave, particularly with digital signals, more than one modulation method can be used on the same carrier. As amplitude and frequency are independent variables, we can simultaneously use AM and FM.

3.2.1 Amplitude Modulation

Amplitude modulation is the process of changing the amplitude of a constant frequency carrier wave in response to changes in the baseband signal. AM is used for transmitting information via a radio carrier wave. In AM, the carrier radio wave amplitude is varied in proportion to the modulating waveform transmitted, e.g. the carrier waveform corresponds to sounds to be reproduced by a loudspeaker. AM was the first modulation method used to transmit voice by radio, starting with Roberto De Moura and Fessenden's radio telephone experiments in 1900 [3.1]. AM is used today in many forms, such as in portable two-way radio, VHF radio and Citizen's Band (CB) radio.

This process is illustrated in figure 3.1 using a sine wave baseband as the signal or modulation.

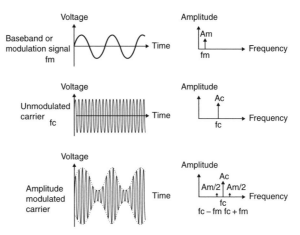

Figure 3.1: *The process of amplitude modulation.*

The modulating baseband m(t), unmodulated carrier and modulated carrier are shown in figure 3.1. It is noted that the original continuous signal is now a 'virtual' signal carried in the peaks of the modulated carrier wave only.

Consider a carrier wave (sine wave) of frequency f_c and amplitude A given by:

$c(t)_{unmod} = A_c \times \sin2\pi\ f_c t$ **(eq 3.2)**

Let $m(t)$ represent the modulating waveform.

We take the modulation to be a sine wave of frequency f_m, a much lower frequency than f_c (such as an audio frequency), so:

$m(t) = A_m \times M \times \cos2\pi\ f_m t$ where M is the modulation index. We chose $M < 1$ so $(1+m(t))$ is positive. If $M > 1$ then overmodulation occurs and transmitted signal reconstruction leads to loss of original signal information.

AM results when the unmodulated carrier $c(t)_{unmod}$ is multiplied by the positive quantity $(1+m(t))$, so:

$c(t)_{mod} = [1 + m(t)] \times c(t)_{unmod}$

$c(t)_{mod} = [1 + A_m \times M \times \cos2\pi\ f_m t]\ \times A_c \times \sin2\pi\ f_c t$

$$c(t)_{mod} = A_c \times \sin2\pi f_c t + A_m \times M \times \cos2\pi f_m t \times A_c \times \sin2\pi f_c t$$

Considering trigonometrical identities or sum and difference frequencies, this expression is the sum of three separate sine waves:

$$c(t)_{mod} = A_c \times \sin2\pi f_c t + \frac{A_m M}{2} \times [\sin2\pi(f_c + f_m)t + \sin2\pi(f_c - f_m)t] \qquad \textbf{(eq 3.3)}$$

Therefore, the modulated signal has three components: an unchanged carrier wave $c(t)_{unmod}$ and two pure sine waves (or sidebands) with frequencies slightly above and slightly below the carrier frequency f_c.

Analysis of the frequency content of the AM carrier in figure 3.1 shows three frequency components: the carrier, the Upper Side Band (USB) frequency and the Lower Side Band (LSB) frequency. In practice, basebands are complex, covering a range of frequencies from the lowest component frequency (f_L) to the highest (f_H). In this form of AM signal, the carrier is accompanied by the USB from ($f_c + f_L$) to ($f_c + f_H$) and the LSB from ($f_c - f_H$) to ($f_c - f_L$).

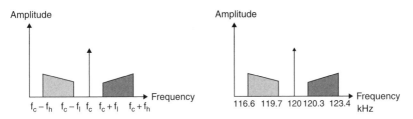

Figure 3.2a: *Generic DSC(FC) A3E transmission. 3.2b: Specific transmission conditions.*

When a radio operator's Press To Talk (PTT) switch on the microphone is operated, a carrier is generated to which speech is added. This type of signal is known as a Double Side Band (Full Carrier) signal, DSB(FC), or technically an A3E transmission, shown in figure 3.2a. We observe the frequency domain diagram of a 120 kHz carrier wave, modulated with speech ranging from 300 Hz to 3.4 kHz, in figure 3.2b.

Signal bandwidth is the frequency range required to contain all the signal frequencies. From figure 3.2a, the highest frequency component of a DSB(FC) signal is $f_c + f_H$ and the lowest is $f_c - f_H$. Hence the bandwidth is the difference between these two components, i.e.

$$BW_{DSB} = (f_C + f_H) - (f_C - f_H) = 2f_H \qquad \textbf{(eq 3.4)}$$

The bandwidth of a DSB(FC) signal is twice the highest information frequency.

Example 3.1: If the electrical base bandwidth of a signal includes frequencies in the range of 2 kHz to 400 kHz, what is the DSB(FC) bandwidth?

DSB(FC) bandwidth $= 2 \times f_H = 2 \times 400$ kHz $= 800$ kHz

---

Intelligible voice shown in figure 3.2b has a range of frequencies from 196.6 kHz to 123.4 kHz, giving a bandwidth of 6.8 kHz, twice the highest baseband frequency (3.4 kHz).

3.2.1.1 AM modulation Index

It is often useful to define the modulation level applied to a signal. A standard factor or *modulation index* M is widely used. The modulation index is important: too low a level of modulation and the modulation doesn't use the carrier efficiently, too high and the carrier can become overmodulated, causing sidebands to extend beyond the allowed bandwidth and resulting in interference to other users.

The AM modulation index and other modulation indices indicate the amount by which the modulated carrier varies around its unmodulated level.

It is expressed as a percentage or a ratio:

$$\textit{Modulation index } M = \frac{A_m}{A_c} \qquad \textbf{(eq 3.5)}$$

where: A_c is the carrier amplitude, and A_m the modulating signal waveform amplitude. Thus M is the modulation index, the peak change in RF amplitude, from its unmodulated value.

Example 3.2: For an AM modulation index with M = 0.5 and an unmodulated carrier amplitude of 1 V ($A_c = 1$ V), what are the *maximum* and *minimum* values of the amplitude modulated carrier wave in terms of peak voltage A_c?

If $M = 0.5 = \dfrac{A_m}{A_c}$, the modulation index causes the signal to increase by a factor of 0.5 and

decrease to 0.5 of its original level, so the carrier will increase to a value of $1\,V + 0.5\,V = 1.5\,V$ peak and have a minimum envelope value of $1\,V - 0.5\,V = 0.5\,V$.

If the modulation depth is 1, (i.e. the carrier and modulation have equal magnitude), the modulation index $M = 1$ and the modulated carrier wave will reach a maximum twice the unmodulated carrier amplitude and a minimum of zero.

$$c(t)_{mod} = A_c \times \sin 2\pi\, f_c t + \frac{A_m M}{2} \times [\sin 2\pi\,(f_c + f_m)\,t + \sin 2\pi\,(f_c - f_m)t]$$

Thus for $M = 1$ $A_c = A_m$ so:

$$c(t)_{mod} = A_c \times \sin 2\pi\, f_c t + \frac{A_c}{2} \times [\sin 2\pi\,(f_c + f_m)\,t + \sin 2\pi\,(f_c - f_m)t]$$

3.2.1.2 Single Sideband (SSB) signals

As HF radio spectrum use increased, it became clear that DSB(FC) transmission wasted available bandwidth as both sidebands contain identical information. The first development removed one sideband from the transmitted signal to produce a Single Sideband (Full Carrier) SSB(FC) signal (or *H3E* transmission), which reduced bandwidth needed but still gives an 'envelope' shaped like the baseband so a signal could be demodulated by a simple detector. The next step partially removed the carrier as well as saving more transmitter power to generate SSB Pilot Carrier (PC) for older equipment to operate that cannot produce an accurate fixed carrier frequency output, resulting in the so-called *R3E* single sideband with reduced carrier.

A modern SSB signal removes the carrier altogether and only transmits one sideband. This means the required transmission bandwidth is the same as the base bandwidth and **all** transmitted power is in the SSB. An advantage of not transmitting the carrier is increased sideband power, but a disadvantage is that as the carrier represents a transmitted reference, this is absent in an SSB Suppressed Carrier SSB(SC) signal or *J3E* transmission. A receiver must generate frequency just as accurately as the original transmitter. With SSB(SC), an error of as little as 100 Hz between original carrier frequency and the one generated in the receiver can make a signal unintelligible. Possible transmission methods for AM are shown in figure 3.3.

Note: Two sideband signals allows transmission of stereo as if the LSB = left + right speaker and USB = left − right speaker; addition and subtraction will result in the pure left and right speaker outputs respectively.

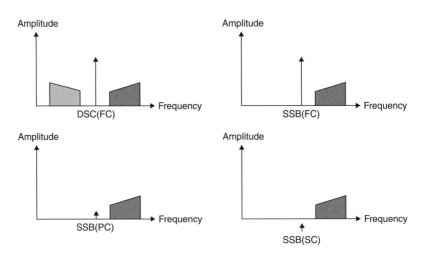

Figure 3.3: *Different AM methods.*

3.2.1.2.1 Power in sidebands

The presence of the two identical sidebands means identical power is 'repeated' in the two sidebands equally, i.e. $P_{LSB} = P_{USB}$.

Now, if: $P_c = (A_c)^2$ and

$$P_{USB} = P_{LSB} = (A_{SB})^2$$

And given: $M = \dfrac{A_m}{A_c}$

Therefore $A_m = M A_c$

so for one sideband with $\dfrac{A_m}{2} = \dfrac{MA_c}{2}$, the power in one sideband

will be: $P_{SB} = \left(\dfrac{MA_c}{2}\right)^2$

And the power in both sidebands $= P_{USB} + P_{LSB}$

$$= \left(\frac{MA_c}{2}\right)^2 + \left(\frac{MA_c}{2}\right)^2 = \frac{M^2}{2}A_c^{\ 2}$$

Thus total power $= P_c + P_{LSB} + P_{LSB} = (A_c)^2 + \dfrac{M^2}{2}A_c^{\ 2}$

So the total power radiated in terms of the total double sideband power leads to:

$$P_T = P_c\left(1 + \frac{M^2}{2}\right) \qquad\qquad \textbf{(eq 3.6)}$$

Example 3.3: If M = 0.3 and the power in the carrier = 1 kW, what is the total power radiated by the system (4 significant figures)?

Using the equation: $P_T = P_c\left(1 + \frac{M^2}{2}\right)$ and substituting for the values gives:

$$P_T = 1000\left(1 + \frac{0.3^2}{2}\right) = 1.045 \text{ kW}$$

Example 3.4: What is the maximum possible power put in before distortion of the 'envelope' of the peaks containing the original waveform takes place?

We need to find the biggest value of M before overmodulation occurs, which is when M = 1.

Hence $P_T = P_c\left(1 + \frac{1^2}{2}\right) = \frac{3}{2}P_c$, of which both sidebands are : $\frac{1}{2}P_c$.

Thus maximum sideband power per sideband is 16.7 per cent of the carrier power, which is a waste in energy if both sidebands are transmitted.

3.3 Angle modulation

Both SSB(SC) and DSB(FC) are AM signals. Since AM signals suffer from vulnerability to amplitude-based noise, which appears as random carrier amplitude variations, noise looks like part of the information to an AM receiver and cannot be removed from the signal without losing some information. The usual method to reduce amplitude-based noise for AM signal is to maintain a high signal S/N. However, as the receiver moves further away from the transmitter, received signal power and thus S/N ratio steadily reduces.

To overcome this noise problem, some form of signal is needed to protect against amplitude-based noise. The obvious solution is to have a form of modulation that doesn't involve changing the carrier amplitude. If the angle term $\sin(2\pi ft + \varphi)$ is used to add baseband information, the amplitude (A_c) of the carrier should stay constant. Angle modulation is immune to amplitude variations caused by noise and can be removed in the receiver as they contain no information. Although both FM and PM can modulate a carrier, it is usual to consider FM with digital basebands.

3.3.1 Frequency Modulation (FM)

FM is the process of changing the frequency of a constant amplitude carrier with the addition of a modulating baseband signal. FM produces a more complex signal than AM on account of the continuous variation of frequency introduced, and consequently occupies a much wider bandwidth. FM was first proposed by Edwin Armstrong, one of the 'founding fathers' of radio technology, who invented the superheterodyne radio receiver in 1918 as well as FM in 1933, overcoming low level signal problems in the presence of noise [3.2].

This increased FM bandwidth requirement means that FM occupies a much wider and higher frequency range than AM, i.e. VHF or SHF, etc., and is essentially LOS transmission. Consequently, maritime users should be aware that when operating on a VHF mobile maritime radio link, a large headland or high cliffs will cause the signal LOS link to be broken. The FM process is illustrated in figure 3.4.

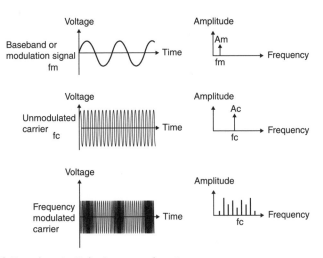

Figure 3.4: *Left: Time domain. Right: Frequency domain.*

FM produces a more complex signal than AM on account of the continuous nature of the frequency variation introduced. In the UK, FM radio stations broadcast between frequencies 88 and 108 MHz, with a channel bandwidth of 200 kHz. FM radio was first deployed in mono in 1940s, while stereo FM was introduced in 1960.

The FM signal frequency spectrum consists of several pairs of sidebands either side of the carrier. The number of pairs and their amplitudes depends on the modulating signal amplitude. It is usual to classify FM signals as *narrow band* if they have one pair of side bands and *wide band* if there is more than one pair. These two types of FM signal are shown (figure 3.5), where the baseband is a single tone of frequency f_m.

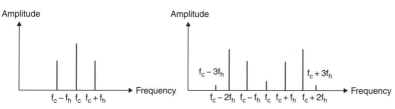

Figure 3.5: *(a) AM and (b) FM spectra.*

The simplest approach to generating an FM signal is to apply the modulation m(t) directly to a Voltage Controlled Oscillator (VCO), as shown in figure 3.6. The signal to be sent helps drive the oscillator output. If the voltage increases, the oscillator frequency increases, and vice versa. An analogy of this FM approach is that of a human heart between rest and under intense exercise; the greater the exercise undertaken, the faster the heartbeat.

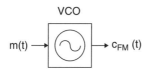

Figure 3.6: *Voltage controlled oscillator.*

If the information to be transmitted (the modulation) is applied to the VCO, the output signal $c_{FM}(t)$ should be a constant amplitude carrier sine wave whose frequency is a linear function of control voltage. When the modulating signal is zero, the carrier wave will be at its centre frequency f_c. When a modulating signal is present, the instantaneous frequency of the output signal will vary above and below the centre frequency and is expressed by:

$$f_i(t) = f_c + K_{VCO} \times m(t) \qquad \textbf{(eq 3.7)}$$

where K_{VCO} is the voltage to frequency gain of the VCO, and $K_{VCO} \times m(t)$ is the Instantaneous Frequency Deviation (IFD). The instantaneous *phase* of the output signal $f_o(t)$ is equal to 2π multiplied by the integral of the IFD, as shown:

$$f_o(t) = 2\pi \, K_{vco} \times \int_0^t m(t)dt$$

The initial phase condition is assumed zero for simplicity.

If the unmodulated carrier takes the form:

$$c(t)_{unmod} = A_c \times \cos 2\pi\, f_c t$$

Then the modulated carrier will take the form:

$$c(t)_{mod} = A_c \times \cos[2\pi\, f_c t + f_o(t)]\text{ , so}$$

$$c(t)_{mod} = A_c \times \cos[2\pi\, f_c t + 2\pi \times K_{VCO} \int_0^t m(t)dt] \qquad \textbf{(eq 3.8)}$$

To estimate the FM signal bandwidth, let us take a single frequency tone signal as used previously for AM, i.e.:

$$m(t) = A_m \times \cos 2\pi\, f_m t\text{, where all quantities are the same as discussed.}$$

We ignore modulation index here by considering M = 1 as we will shortly consider a separate FM modulation index.

Substituting this modulation signal into the above formula, we find:

$$c(t)_{modulated} = A_c \times \cos[2\pi\, f_c t + 2\pi \times K_{VCO} \int_0^t A_m \times \cos 2\pi\, f_m t\, dt]$$

And so:

$$c(t)_{mod} = A_c \times \cos[2\pi\, f_c t + 2\pi \frac{K_{VCO} \times A_m}{f_m} \times \sin 2\pi\, f_m t]$$

$$c(t)_{mod} = A_c \times \cos[2\pi\, f_c t + 2\pi \frac{\Delta f}{f_m} \times \sin 2\pi\, f_m t]$$

$$c(t)_{mod} = A_c \times \cos[2\pi\, f_c t + 2\pi\beta \times \sin 2\pi\, f_m t] \qquad \textbf{(eq 3.9)}$$

$K_{VCO} \times A_m$ represents the peak frequency deviation of the FM signal from the centre frequency and is directly proportional to the amplitude of the modulating signal and the VCO gain. The quantity Δf is termed the maximum instantaneous frequency deviation. The ratio of the frequency deviation Δf to the modulation signal frequency f_m is called the modulation index, β. For a single tone modulating signal, the number of significant sidebands in the output spectrum is a function of the modulation index. This can be seen by writing the FM output signal in terms of the nth order Bessel functions of the first kind:

$$c(t)_{mod\ FM} = A_c \sum_{n=-\infty}^{\infty} J_n(\beta) \times \cos[2\pi\ (f_c t + nf_m)t] \qquad \textbf{(eq 3.10)}$$

Further details about the important role played by the Bessel function can be found elsewhere [3.3–3.4].

The equally spaced harmonic distribution of a sine wave carrier modulated by a sinusoidal signal is represented with Bessel functions, which provide the basis for a mathematical understanding of FM in the frequency domain. The magnitude of the respective equally spaced sidebands produced have values corresponding to their corresponding Bessel functions output as a function of β. The number of sidebands of an FM signal and its associated magnitude coefficient are found with the help of Bessel function tables such as the one shown in Table 3.1.

β	J0	J1	J2	J3	J4	J5	J6	J7
0	1							
0.25	0.98	0.12						
0.5	0.94	0.24	0.03					
1	0.77	0.44	0.11	0.02				
1.5	0.51	0.56	0.23	0.06	0.01			
2	0.22	0.58	0.35	0.13	0.03			
2.5	−0.05	0.5	0.45	0.22	0.07	0.02	0.01	
3	−0.26	0.34	0.49	0.31	0.13	0.04	0.01	
4	−0.4	−0.07	0.36	0.43	0.28	0.13	0.05	0.02

Table 3.1: *Bessel functions of the first kind, rounded to two decimal places.*

By taking the Fourier transform, we obtain the discrete FM output spectrum with magnitude coefficients as a function of β.

$$c(f)_{mod\ FM} = A_c \sum_{n=-\infty}^{\infty} J_n(\beta) \times \cos[\delta\ (f - f_c - nf_m) + \cos[\delta\ (f + f_c + nf_m)]$$

3.3.2 Bessel functions

Bessel's equations are useful in cylindrical or spherical coordinates, which occur surprisingly often. For example, tap a water-filled cup and you observe Bessel functions, like a sine wave but one decaying rapidly in amplitude with distance from the centre of the water cup surface (figure 3.7, see plate section).

The similarity with the sine wave distribution is possibly more easily noted when considering the earlier equation modulation integral $\int_0^t m(t)dt$ for a modulation of $m(t) = A_m \times \cos2\pi f_m t$ so that: $\int_0^t A_m \times \cos2\pi f_m t \ dt = \dfrac{A_m \times \sin2\pi f_m t}{2\pi f_m}$ which is of the form $\dfrac{\sin x}{x}$ characteristic of the Bessel function, decaying increasingly in amplitude for increasing x. Various Bessel functions, e.g. $J_0(x)$, look like decaying sine waves (figure 3.8).

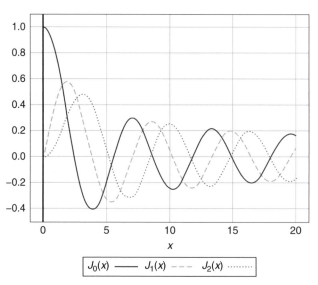

Figure 3.8: *Plot of Bessel functions of the first kind, $J_a(x)$, for integer orders $a = 0, 1, 2$.*

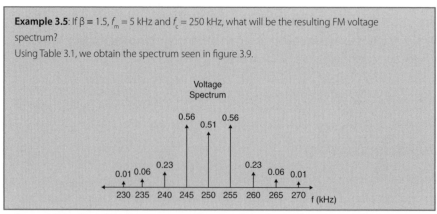

Example 3.5: If $\beta = 1.5$, $f_m = 5$ kHz and $f_c = 250$ kHz, what will be the resulting FM voltage spectrum?

Using Table 3.1, we obtain the spectrum seen in figure 3.9.

Figure 3.9: *Resulting FM voltage spectrum for example 3.5, $\beta = 1.5$, $f_m = 5$ kHz and $f_c = 100$ kHz.*

Note: As the FM modulation index β changes, existing sidebands may decrease and increase in size, while further sidebands start to 'pop up' in the fringes of the FM bandwidth.

The FM modulation index β determines the signal bandwidth by determining the number of effective FM signal sidebands. For instance, if $\beta = 0.25$, only one sideband is needed, however if $\beta = 1$, three sidebands are necessary. Another important point about the FM modulation index is that it changes a lot even for *fixed* frequency deviation because the modulation frequency can vary. In general, as the FM modulation index increases, the number of sidebands increases and bandwidth increases. If β is much less than 1, the modulation is called *narrowband FM*, and its bandwidth is approximately twice the modulation frequency.

With a tone-modulated FM wave, if the modulation frequency is held constant and the modulation index is increased, the (non-negligible) FM signal bandwidth increases but the spacing between spectra stays the same; some spectral components decrease in strength as others increase. If the frequency deviation is held constant and modulation frequency increased, the spacing between spectra increases.

Narrowband FM is used for two-way radio systems where the carrier deviates only 2.5 kHz above and below the centre frequency with speech signals of no more than 3.5 kHz bandwidth. Wideband FM is used for broadcasting in which music and speech are transmitted with up to 75 kHz deviation from the centre frequency and can carry audio frequencies up to 20 kHz bandwidth. For some values of modulation index, carrier amplitude becomes zero and all signal power is in the sidebands.

For a carrier modulated by a single sine wave, the resulting frequency spectrum can be calculated using Bessel functions of the first kind, as a function of sideband number and modulation index. Suppose we limit ourselves to only those sidebands that have a relative amplitude of at least 0.01: as modulation level increases, smaller ripples in frequency become apparent about both higher and lower values than the nominal unmodulated carrier frequency, so FM exhibits not only increasing sideband size with increasing modulation depth but an increase in the number of sidebands as well.

The characteristics of AM and FM are compared in table 3.3.

	Amplitude Modulation AM	Frequency Modulation FM
Change A$_m$	1. Bandwidth is unchanged. 2. The size of the sidebands is changed. 3. The number of sidebands is unchanged.	1. Bandwidth changed as the number of sidebands changes. 2. Size of sidebands varies with the number present.
Change f$_m$	1. The bandwidth is changed. 2. The size of sidebands is unchanged. 3. The number of sidebands is unchanged.	1. The bandwidth is changed. 2. The size of sidebands is unchanged. 3. The number of sidebands is unchanged.

Table 3.3: *AM vs FM.*

3.4 Radio receiver principles

3.4.1 Receivers

A receiver is a device that:

1) Selects a wanted signal from any unwanted signals, and;

2) Retrieves the modulation information contained in the detected signal.

This is a useful definition as it shows the importance of a receiver's selection stage, whether radio, radar or acoustic sonar receivers. The main difference between the two electromagnetic uses is that in radio, unwanted signals are noise and other radio signals, while in radar the unwanted signals are noise and clutter (Chapter 6). In either case, it is important that a receiver selects the wanted signal and rejects all others.

3.4.2 Selectivity

A receiver is considered as a *matched* filter, designed to ensure the maximum transfer of signal to noise ratio from input to output. A matched filter defines the effective receiver bandwidth. If the receiver's selecting stage bandwidth is too narrow, part of the signal is lost, yielding a lower S/N ratio. However, if a receiver's bandwidth is too wide, extra noise enters the receiver and the S/N ratio is reduced. Observe that for each signal there is an optimum receiver bandwidth, decided by the signal's bandwidth (figure 3.10).

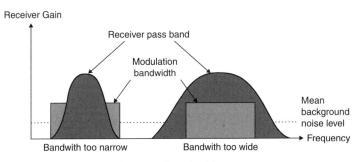

Figure 3.10: *Receiver selectivity as a function of bandwidth.*

For good selectivity, the selection frequency response ideally has a flat top and steep, preferably vertical, sides. In practice, the filter response has a shape similar to that shown (figure 1.19, 1.20) and some method must be used to assess the sharpness of the cut-off points of the filter's response. A common method is to specify a – 50 dBs (1:10^5 ratio) as well.

3.4.3 Sensitivity

Sensitivity is critical to signal information retrieval. The processes involved here are amplification of a weak received signal and use of a suitable detector to recover the information. Receiver sensitivity is defined as the *minimum* input signal that gives the full analogue output with adequate S/N ratio.

But what comprises an adequate S/N ratio?

This depends on the type of information in the signal and its vulnerability to noise interference. Receiver sensitivity is partially a measure of receiver gain, which is also decided by the output S/N ratio.

Even an ideal receiver has a sensitivity limit, as it will receive and amplify noise as well as signal energy. If the signal is less than the noise, the signal cannot be read. All real receivers introduce some noise into the system and this further limits sensitivity.

3.4.4 The basic superheterodyne (superhet) receiver

Figure 3.11 shows the basic block diagram of a superhet radio receiver. Each individual block's primary purpose is shown in table 3.2, and then described in detail.

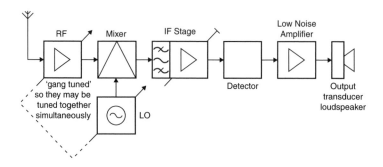

Figure 3.11: *Block diagram of basic superheterodyne radio receiver.*

Block(s)	Purpose(s)
Radio Frequency (RF) amplifier	1. Improves S/N and provide a low noise front end to the receiver.
	2. Rejects the image channel.
	3. Prevents Intermediate Frequency (IF) breakthrough.
	4. Prevents local oscillator reradiation.
Local oscillator and mixer	Translates the wanted signal frequency to the IF.
Domestic superhet Image Frequency (IF) stage AM IF = 0.47 MHz FM IF = 10.7 MHz International standards	1. Selects the wanted signal and rejects all the others. 2. Amplifies the modulation signal to drive the detector.
Detector	Recovers the information from the modulation signal.
Low noise audio amplifier	Amplifies the 'recovered' information to drive the output transducer.
Output transducer	Converts the information into a form a human can understand, e.g. the loudspeaker.

Table 3.2: *Block components of a superheterodyne radio receiver.*

The RF amplifier overcomes a deficiency in the mixing process that produces both a wanted frequency and an unwanted Image Frequency (IF) for each radio signal. This means that, on reception, a similar mixed unwanted image frequency and the wanted RF can be amplified. The RF amplifier is a bandpass filter tuned to the wanted frequency and rejects the image frequency, so:

$$f_{Image} - f_{Lo} = f_{IF}$$ (**eq 3.11**)

as well as

$$f_{Lo} - f_{RF} = f_{IF}$$ (eq 3.12)

Thus: $f_{Image} - f_{Lo} = f_{Lo} - f_{RF}$

$$f_{Image} - f_{Lo} = f_{Lo} - f_{RF}$$ (eq 3.13)

$$f_{Image} = 2f_{LO} - f_{RF}$$ (eq 3.14)

Example 3.6: If the f_{IF} = 10.7MHz and the f_{RF} = 450 MHz what is the f_{LO} and the image frequency (4 significant figures)?

Using: $f_{LO} - f_{RF} = f_{IF}$

So: $f_{LO} = f_{IF} + f_{RF} = 10.7 + 450 = 460.7$ MHz

Using : $f_{Image} = 2f_{LO} - f_{RF} = 2 \times 460.7 - 450 = 471.4$ Hz

---

The RF amplifier helps reduce noise entering the receiver by restricting bandwidth, and by amplifying the RF signal to overcome losses in the mixer.

The mixer multiplies the local oscillator output with the RF amplifier output to produce the sum $f_c + f_m$ and difference $f_c - f_m$ frequencies. The receiver is tuned by altering the Local Oscillator (LO) frequency until the difference frequency equals the tuned IF amplifier frequency. The IF stage is a fixed frequency bandpass amplifier, whose bandwidth should match the wanted modulation signal, providing most of the receiver amplification and determining the final detected S/N ratio. The detector extracts information that has been placed upon the carrier wave. For radio, an audio frequency amplifier amplifies the very low detector output power so that it can drive the chosen output transducer.

Consider the reception conditions that permit two possible values of the LO frequency, and the image frequency that results in each case.

Example 3.7: A signal frequency of 3 MHz is fed into the mixer of a superheterodyne radio receiver with an intermediate frequency of 600 kHz. Calculate (i) the *two* possible values of the local oscillator frequency and (ii) the two image frequencies that result.

Consider the antenna structure shown in figure 3.12:

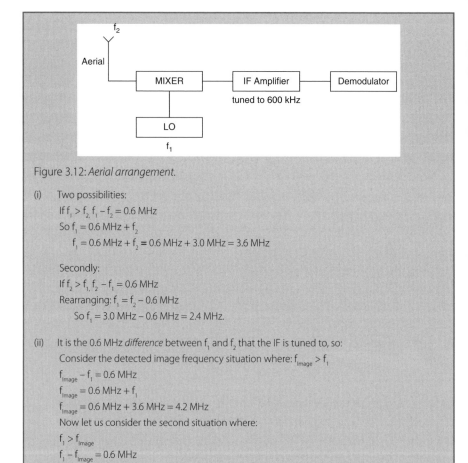

Figure 3.12: *Aerial arrangement.*

(i)   Two possibilities:
       If $f_1 > f_2, f_1 - f_2 = 0.6$ MHz
       So $f_1 = 0.6$ MHz $+ f_2$
           $f_1 = 0.6$ MHz $+ f_2 = 0.6$ MHz $+ 3.0$ MHz $= 3.6$ MHz

       Secondly:
       If $f_2 > f_1, f_2 - f_1 = 0.6$ MHz
       Rearranging: $f_1 = f_2 - 0.6$ MHz
           So $f_1 = 3.0$ MHz $- 0.6$ MHz $= 2.4$ MHz.

(ii)   It is the 0.6 MHz *difference* between f_1 and f_2 that the IF is tuned to, so:
       Consider the detected image frequency situation where: $f_{Image} > f_1$
       $f_{Image} - f_1 = 0.6$ MHz
       $f_{Image} = 0.6$ MHz $+ f_1$
       $f_{Image} = 0.6$ MHz $+ 3.6$ MHz $= 4.2$ MHz
       Now let us consider the second situation where:
       $f_1 > f_{Image}$
       $f_1 - f_{Image} = 0.6$ MHz
       $f_{Image} = f_1 - 0.6$ MHz
       $f_{Image} = 2.4$ MHz $- 0.6$ MHz $= 1.8$ MHz

Self-assessment questions

3.1   Explain why modulation is necessary for practical radio and radar transmission. If an electrical baseband has frequencies 550 kHz to 650 kHz, what is the electrical base bandwidth (1 significant figure)?

3.2   If a broadband aerial is used with centre frequency $f = 600$ kHz and $Q = 75$, what is the ideal dipole length (2 decimal places)?

3.3   Explain what is meant by AM and FM, illustrating your answer with diagrams. What are the key differences between AM and FM?

3.4   State the bandwidth of a DSB(FC) signal. Explain the advantages and disadvantages of SSB(SC) compared with DSB(FC) transmission. What do you understand by the following transmission modes: A3E, H3E, J3E, R3E?

3.5   Prove the theory of AM modulation.

3.6   Explain why FM is better than AM at producing high signal to noise ratios in the receiver.

3.7   Contrast the variations in bandwidth of AM and FM signals and the typical carrier frequencies required.

3.8   Draw the block diagram of a superheterodyne radio receiver.

3.9   Explain the purpose of each block in the superheterodyne radio receiver in question 3.8.

3.10   A superheterodyne radio receiver has its aerial connected directly to the mixer. The LO is set at 1.50 MHz and the IF amplifier operates at 500 kHz and has a bandwidth of 20 kHz. Four signals of frequencies 1.00, 1.01, 1.02 and 2.00 MHz are received at the mixer.

  (i) What frequencies are produced at the mixer output?

  (ii) What frequencies are produced at the IF amplifier output?

  (iii) What is the effect of placing a tuned circuit of central frequency 1.01 MHz and bandwidth 40 kHz between the aerial and the mixer?

  (iv) What is the effect of *decreasing* the IF bandwidth to 10 kHz?

REFERENCES

[3.1] 'Recent Progress in Wireless Telephony', Reginald A Fessenden, *Scientific American*, Vol. 96 no. 3. (19 January 1907), pp. 68–69 (retrieved 10 February 2016).

[3.2] users.erols.com/oldradio/

[3.3] *Communication Systems 3rd edition*, S Haykin (Wiley, 1994, 978–0–4715–7176–6).

[3.4] *Principles of Communications, Systems, Modulation, and Noise 4th edition*, RE Ziemer and WH Tranter (Wiley, 1995, ISBN 978–0–4711–2496–2).

4

Digital Signalling Methods and Digital Keying

'Simply let your "Yes" be "Yes", and "No", "No".'   Matthew 5:37

4.1 The bandwidth of a digital signal

Digital basebands were introduced in Chapter 1. Unlike analogue signals (with a clearly defined bandwidth between upper and lower frequencies), digital signals contain a very wide (theoretically infinite) range of frequencies, significantly wider than for analogue FM. This makes it impossible to specify an exact 'modulation' bandwidth. The approach taken for real communication channel and receiver design is aimed at finding a practical minimum signal bandwidth that allows the digital signal to be accurately identified in a digital receiver. To obtain the minimum bandwidth, we must consider the range of possible digital signals transmitted, ranging from a signal containing all zeros through to one containing all ones. For an 8 bit sample, 11111111s and 00000000s are both steady signals with zero frequency (no change). The most rapid variation appearing in a digital signal is a sequence of alternating ones and zeros, i.e. 10101010 or 01010101.

The minimum bandwidth must allow a receiver to follow this repetitively alternating variation (figure 4.1).

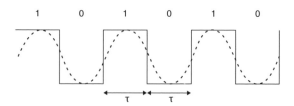

Figure 4.1: *Most rapid digital waveform sequence.*

Looking at figure 4.1 one sees the lowest frequency sine wave that has an alternate 101010101 variation (shown by the dotted line), with a periodic time of 2τ and frequency

$$f = 1/T = \frac{1}{2\tau}$$ **(eq 4.1)**

If the receiver operates with this cosine wave, and sets the fastest variation, this will be the minimum frequency the system must pass for the receiver to follow the 10101010 variation. The lowest frequency is found by looking at the two extremes of all ones or all zeros. It is clear the period of the minimum sine wave that describes the repetitive sequence is the digital '10' period. A square wave is constructed from the minimum frequency and a sequence of harmonically related frequencies (frequencies at multiples of this lowest minimum frequency), and thus a digital waveform requires *more* frequencies (and therefore bandwidth) than its equivalent sine wave. Generation of the square wave, by addition of harmonic frequencies, is given in the left of figure 4.2. First the fundamental frequency is given, then a much smaller amplitude of the first harmonic is added, then the second harmonic, and the third harmonic, etc.

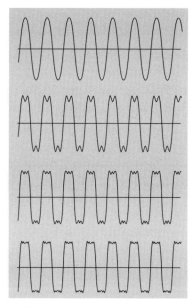

Figure 4.2: *Fourier construction of a digital square wave using a sequence of harmonically related frequencies.*

Digital signal bandwidth, as with analogue, is taken to be the difference between the lowest and highest frequencies, so the minimum bandwidth required for a digital signal is:

Minimum bandwidth (B) = Highest frequency – lowest frequency

$$= \frac{1}{2\tau} - 0$$

$$= \frac{1}{2\tau}$$

The bit rate of a digital signal- is given by $\frac{1}{\tau}$ and it is usual to specify the minimum bandwidth in terms of its bit rate, so:

Minimum bandwidh $B = \dfrac{bit\text{-}rate}{2}$ (**eq 4.2**)

Example 4.1: If the bit rate is 3000 bits per second, what is the minimum bandwidth (2 significant figures)?

Using $B = \dfrac{bit\text{-}rate}{2} = \dfrac{3000}{2} = 1500$ Hz

---

Analysis of the figure 4.2 waveform shows a digital signal is *not* a single sine wave but is made of different harmonically related frequencies. Digital signals differ from complex analogue signals as the component frequencies are all harmonically related to the fundamental digital signal frequency, itself derived from the period of the square wave of duration T.

4.2 Capacity of a channel

The bandwidth requirement of a digital signal is related to the rate at which bits, and thus digital data, are transmitted. Speed in a digital system is measured in bits per second (bps). For a digital communications channel, the maximum data rate (baud speed or capacity) obtained from the above equation is:

Maximum data rate = 2 B   (**eq 4.3**)

where B is the channel bandwidth. Baud speed is the signalling speed in state transitions per second and B the channel bandwidth. For a simple binary digital transmission system, baud speed is equal to bits per second, but for more advanced systems the bit rate is greater than the baud speed.

> **Example 4.2:** If the bandwidth of a TV signal is 50 MHz, what is the maximum date rate possible?
>
> Using the capacity of a channel results in a Maximum data rate = 2 B
>
> Maximum data rate = 2 × 50 MHz = 100 M bits per second.

4.3 Capacity of a noisy channel

The presence of noise in a real channel reduces the channel's ability (and capacity) to carry as much information as one without noise. The capacity of a noisy channel is given by the Shannon-Hartley theorem, which states:

$$Maximum\ data\ rate = 2B \times log_2\left(1 + \frac{S}{N}\right) \qquad \textbf{(eq 4.4)}$$

This practical law gives the maximum capacity of a noisy channel of given bandwidth and S/N ratio. In practice, the capacity of a noisy channel is *less* than this theoretical figure. Some key applications of this law will be introduced later when discussing spread spectrum techniques.

> **Example 4.3:** If the S/N ratio is 2 and the channel bandwidth is 20 kHz, what is the maximum data rate possible on the channel (5 significant figures)?
>
> Use: $Maximum\ data\ rate = 2B \times log_2\left(1 + \frac{S}{N}\right)$ and
>
> substituting for the values given:
>
> $Maximum\ data\ rate = 2 \times 20 \times 10^3 \times log_2(1 + 2)$
> $\qquad\qquad\qquad = 4 \times 10^4 \times log_2(3)$
> $\qquad\qquad\qquad = 19084\ bits\ per\ second$

4.4 Advantages of digital signalling

Modern communications systems use digital signalling as it offers advantages over conventional analogue signalling. This is true of maritime communications, and is generally true of civilian and domestic systems. For instance, analogue sound reproduction systems of vinyl records and magnetic tape are now largely replaced by digital Compact Discs (CDs) for music and Digital Versatile Discs (DVDs) for video. Digital radio and television are now available throughout large parts of the world.

Clearly there are good reasons for the transition to digital signalling, and it is these reasons that we will examine.

Digital signals are commonly produced as binary signals with two states. We have seen that AM signals suffer from noise degradation as they travel through a communications channel. In the case of binary digital, information is encoded as two voltages. As long as the separation of these voltages is large compared with the noise, it is easy to read which voltage state (high or low) a signal is in. Unlike analogue signals, a digital system doesn't read the exact voltage, but only which *state* a signal is in. This gives binary signals considerable advantages over AM in terms of error-free transmission.

Figure 4.3 shows that, despite the significant noise modulation, the two states of a binary signal are still clearly distinguished and thus totally separated. The result is loss-free extraction of the original digital signal. Remember, amplitude variations produced by the noise would be read as part of the information by analogue communication systems.

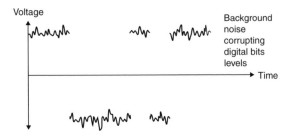

Figure 4.3: *Digital transmission with noise added.*

As well as creating more accurate signals, digital signalling produces other advantages. On channels prone to errors, particularly radio channels, extra parity binary bits may be used to protect data from corruption. Digital signals are also protected against unauthorised reception by encrypting them, so only the authorised users access the encrypted data. The processes of encryption and decryption are dealt with later, when we consider security (section 4.6.4).

4.5 Error Detection and Correction (EDAC)

The two main classes of EDAC are Automatic Request for Repetition (ARQ) and Forward Error Correction (FEC). Both methods involve extra parity bits. In the case of ARQ, parity bits allow a receiver to detect an error and request retransmission. FEC systems use a logical process on the received data to identify an error and correct it. EDAC techniques can protect data bits from corruption during transmission. FEC is valuable in situations where request for retransmission of information may take too much time, e.g. in a fire-control radar system requesting search parameters to be sent again, or is impossible, as from a space probe destroyed in the process of data acquisition, e.g. Huygen's space probe sent to Titan's thick crushing atmosphere.

4.6 Analogue to digital conversion

Discussion of digital signals would be incomplete without some explanation of how digital signals are produced and how their information is recovered. To take advantage of digital signalling methods, we need a digital baseband. Although nearly all computer-generated basebands are digital, some of the most common basebands used by humans are analogue. Human basebands include voice, music and most types of positional information and display. To use digital signalling techniques with these essential analogue basebands, analogue basebands must be converted into a digital format. This process is called *analogue to digital conversion* (or A to D conversion). We will look at the most common methods for digitising analogue basebands.

The process involves typically three or four processes:

1. Sampling: obtaining the frequency of the sampling required.

2. Quantisation: measuring sample voltage amplitude.

3. Encoding: samples are converted into a series of ones and zeroes.

4. Encryption: making data secure (not all systems do this, for varied reasons).

4.6.1 Sampling

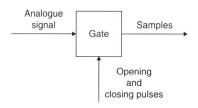

Figure 4.4: *Digital sample gating.*

Sampling is the first step taken in our analogue to digital conversion (figure 4.4). Sampling is considered as a continuous analogue signal that enters the 'gate', which is opened for a short time, and the signal recorded at this sampling time τ, and closed to be reopened a time T seconds later. These pulses are considered as a '1' when the gate is open and a '0' when the gate is closed. Essentially, the process of sampling is a *multiplication* of analogue voltage over the time τ by the unit amplitude gating pulse. Sampling needs to take place sufficiently fast to 'capture' the most rapid variations in a signal that is being sampled (figure 4.5).

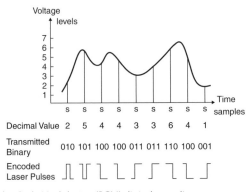

Figure 4.5: *Simple Pulse Code Modulation (PCM) digital encoding.*

For accurate analogue baseband recovery, the minimum sampling frequency must be *at least* twice the highest frequency of the analogue baseband. From a practical engineering view, to establish a common protocol of how fast to sample it is usual to apply a 10 per cent margin of error to ensure sufficient separation, so:

practical sampling frequency $= 2.2\ f_H$ (**eq 4.5**),

where f_H is the frequency of modulation.

This relationship between the sampling frequency and the highest frequency component of the analogue baseband was first derived theoretically by the Swedish-born American electrical and communications engineer Harry Nyquist (1889–1976) [4.1].

The resulting equation is called the *Nyquist sampling theorem*, or simply 'the sampling theorem'. The rate at which samples are taken is controlled by the sampling theorem, which states that the sampling frequency must be at least twice the highest information frequency. In practice, a higher sampling frequency, at a value of $2.2 \times f_H$, enables easier decoding in the receiver and provides a suitably accurate horizontal series of grid lines. The number of binary digits used to encode the sample controls the accuracy. Careful selection of the number of bits and sampling frequency enables an exact copy of the original signal to be extracted from the encoded signal.

Example 4.4: If the highest frequency in an analogue video signal is 4 MHz, what is the necessary sampling frequency?

Sampling frequency $= 2.2 \times f_H = 2.2 \times 4$ MHz $= 8.8$ million samples per second.

4.6.2 Quantisation

In the previous diagram, regular spaced sampling times are marked with an 's' when the amplitude of the analogue signal is read. Since a digital signal cannot represent a fraction, the voltage in mV is read to the nearest integer with a fraction ignored, e. g. 5.2 V is read as 5 V.

The 'error' between the instantaneous value and the recorded discrete value is the *quantisation error*. The key to a good digital copy of, say, music is to ensure the quantisation error is sufficiently small so the human ear cannot distinguish the measuring error introduced into the signal.

For example, a voltage sample (5) is converted into a binary number, in this case a 3-bit binary sequence (101), and a binary output voltage signal is generated from it for optical fibre communications in the form of a sequence of on and off laser pulses. The resulting stream of binary number groups is called a *Pulse Code Modulated* (PCM) signal. Synchronising signals are also required to enable the receiver to correctly identify each group and decode it back again into a voltage.

Now, 0.2 V represents a rather large fraction to ignore. To increase accuracy, smaller intervals can be used, such as millivolts, so the above voltage would read 5200 mV, but this requires further vertical levels or lines.

4.6.3 Encoding

Encoding requires more than three binary digits for representation in a modern communications system. In practice, 32, 64 and 128 bits are used. Consider one common digital method, Pulse Code Modulation (PCM), used to digitally represent sampled analogue signals. In a PCM sample stream, the vertical amplitude of the analogue signal is sampled regularly at uniform intervals, and each sample quantised to the nearest value within a range of digital steps.

Example 4.5: What is 5200 represented in 16-bit binary?

5200 is equivalent to 0001010001010000. Base 10 to binary conversion is available at an easily accessible website [4.2].

But why is the number of bit sequences *so* long for digital samples? Human beings are used to thinking of numbers in base 10. The symbols used are 0, 1, 2, 3, 4, 5, 6, 7, 8 and 9 – a limited set of symbols compared with those used in written languages. Fortunately, the inherent human anthropomorphic feature of ten fingers has driven counting in base 10 to a common standard globally. Base 10, like all number systems including digital, is based on a place value system, meaning the value of a digit depends on its position within the number.

Now, a problem arises in base 10, and that is: what do we do when we want to represent 'one more than 9', as 9 is the largest symbol we have? The answer is that we will have to reuse the existing symbols to represent larger numbers. Of course, in reality the solution chosen is to consider a new column of numbers to the *left* of the first column representing groups of '1 more than 9' represented *symbolically* with the symbols 10 or 'ten'.

This now allows us to count up to 99 without problem, and represents 9 groups of 'one more than 9' and one group of 9 in the 'units' column on the right. However, to represent 'one more than 99' is again impossible without addition of further symbols, and so a third column is added to count groups of 'one more than 99' (the lowest representation we can achieve using three columns in this way will be one group of 'one more than 99', and no groups of '1 more than 9' and no units, represented as 100 and expressed in English as 'one hundred').

This may appear a laboured point, but consider what happens when we have a digital signal. In the case of a digital signal, we have only two possible states or symbols: a one '1' and a zero '0'. Hence, the largest value we can represent in the unit column is a one! Thus, to record a value one greater than one (i.e. 2) we need a new column to the left of the units column that represents a 2 (this will appear as 10 in binary notation). This similar structuring of binary and base 10 counting means that, reading right to left, the first number represents a 0 and the second number a 2.

It is easy then to represent a 3 with two columns of numbers as this is a combination of 2 + 1, represented by 11. However, we again find that to record one more than this with two columns isn't possible and we need a third column to represent a group of 4, so 4 is represented by 100.

Clearly, we have used three columns already due to our lack of different binary symbols, whereas we haven't even used half of our available base 10 symbols and can still write increasing numbers in the first base 10 unit column.

Example 4.6: Show from first principles how 13 can be represented by a 3-bit data sequence and what it will be.

Consider a 3-bit sequence: the largest column can 'hold' a maximum group of 8, so can we subtract a group of 8 from 13 to fill the column with a 1?

We can, since 13 – 8 is 1 group of 8, leaving a remainder of 5

and 5 – 4 is 1 group of 4, leaving a remainder of 1

and 1 – 1 is a group of 1, leaving 0 – no remainder.

Equivalent to 111 in binary.

Now, for an 8 level system (0 to 7 voltage levels inclusive) requires three bits, while a 16 level system requires four bits, etc.

On a general basis, the number of levels required = 2^N (**eq 4.6**),

where N is the number of bits.

Example 4.7: If we transmit a digital signal with eight bits, how many levels are available?

The number of levels required = $2^N = 2^8 = 256$

4.6.4 Encryption

In cryptography, encryption is the process of encoding information so that only authorised users can read it, while unauthorised users cannot [4.3]. Encryption does not necessarily prevent possible interception, but denies access to the message contents, and the use of special codes to enable spread spectrum can mean interception or deception is prevented.

4.6.4.1 Cryptography

In some civilian and most military communications, there is a fundamental need to keep the contents of signal traffic secure from unauthorised readers. There are many types of codes that can be used to give security to signal traffic. We will confine ourselves here to examining online encryption techniques employing a *stream cypher*. Typically, this is a cryptographic system that gives long term security to digital signals and is the most commonly employed technique [4.4].

Encryption has been used by military and governments to facilitate secure communications. It is also now commonly used to protect information within different kinds of civilian systems, including banking and company data. Encryption is also used to protect data in transit, for data transferred via networks e.g. the internet, mobile telephones, wireless microphones, Bluetooth devices (discussed later) and bank ATM machines.

One of the most secure cryptographic techniques is a One-Time Pad (OTP). This is a pad of cyphers, each used to encode one message only. As each cypher is used once, there is no possibility of an unauthorised listener being able to build up a database over time to assist decryption. This gives the one-time pad a very high degree of security. A stream cypher is an automated form of one-time pad and obtains the same high degree of security if it does not repeat during its operational life. A daily changing stream cypher must not, therefore, repeat during its 24-hour operational life.

At the heart of such systems is a simple process called *Modulo-2 addition*.

4.6.4.2 Modulo-2 addition

Modulo-2 addition is a form of binary addition with no carry. The 'truth' or logic table for this Modulo-2 addition is shown in table 4.1.

A	B	Output
0	0	0
0	1	1
1	0	1
1	1	0

Table 4.1: *Truth table for Modulo-2 addition.*

4.6.4.3 Pseudo Random Numbers (PRNs)

At the heart of stream cyphers lies the production of a Pseudo Random Number Sequence (PRNS). The device that generates the PRNS is the Pseudo Random Number Generator, or PRANG for short. A PRNS is a number sequence that looks as though it were a random number, but has been produced deliberately. A truly random number stream cannot be used in cryptography as it is necessary to produce the same number sequences during both the encryption and decryption processes.

4.6.4.4 The encryption process

The PRNS generated by the PRANG is referred to as the *key stream* and the exact form of the key stream is determined by the *keymat*, the start information for the PRANG and changed for each operating cycle (daily or weekly as required). A *keymat* (keying material) is a generic term for any (usually classified) material used in the process of loading keys into a cryptographic device. A key stream is then modulo-2 added to the digital signal. Physical delivery of civilian keying material must be distributed securely, as is military material for a predetermined period of time.

The process of encryption using a stream cypher is the *synchronous* modulo-2 addition of the key stream to the sample text, or *plain text*. After encryption, the secured text is the *cypher text*, which becomes the transmitted message. The encryption process is illustrated (figure 4.6).

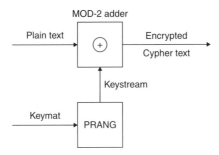

Figure 4.6: *The encryption process.*

The sequence of events can be seen below for an 8-bit binary sequence:

01011001	Plain text	(Data to be sent)
11110100	Key stream	(Encrypting code)
---------------- Modulo-2 addition		
10101101	Cypher text	(Transmitted secure data)

At the receiver, an identical PRANG, using the same keymat, generates a synchronised key stream. This is Modulo-2 added to the received cypher text to recover the plain text.

10101101	Cypher text	(Secured received data)
11110100	Key stream	(Decrypting code)
---------------- Modulo-2 addition		
01011001	Recovered plain text	

The beauty of this method is that we can check if we have recovered the correct plain text as the two digital sequences at the start and the end should be identical.

We have not explicitly discussed *synchronisation*, but without synchronisation of the two data streams it isn't possible to obtain the correct plain text output. For example, an error of 1 bit duration in time results in the wrong output sequence:

10101101	Cypher text (received)
01111010	Key stream misaligned by 1 bit to the right
11010111	Wrong recovered plain text

It is possible to decrypt the encoded information without possessing the original encryption key, but for a well-designed encryption scheme, composed of many bits, a very large computational resource is required. An authorised user can easily decrypt the message with the key provided by the transmitting station.

Various symmetric key encryption methods are used, such as public key encryption, where the encryption and decryption keys are identical (as above), where communicating parties must use the same key before they can achieve secure communication. In public encryption, only the receiving path has access to the decryption key that enables messages to be read [4.5].

4.7 Digital bit rate and bandwidth

A signal using millivolts requires a much higher bit rate than one using volts. The final PCM signal bit rate depends on both sampling frequency (f_s) and the number of bits used to encode each sample (N).

The bit rate is given by multiplying each sample with the number of bits for every sample transmitted every second:

$f_s \times N$ (**eq 4.7**)

The minimum bandwidth $= \dfrac{Bit\ rate}{2}$ so that the resulting minimum bandwidth of

the PCM signal: $\dfrac{Bit\ rate}{2} = \dfrac{f_s \times N}{2}$ (**eq 4.8**)

Example 4.8: For a PCM signal requiring 256 levels and a sampling frequency of 64k bit/s, what is the minimum bandwidth required (3 significant figures)?

The number of levels required $= 2^N$

So if 256 levels are required: $256 = 2^N$

So $N = 8$

Using the equation for the minimum bandwidth $Bandwidth = \dfrac{f_s \times N}{2}$

$Bandwidth = \dfrac{f_s \times N}{2} = \dfrac{6400 \times 8}{2} = 25.6 \ kHz$

The universal standard for voice digitisation is PCM at a transmission rate of 64k bits/s. This results from a sampling rate of 8000 samples per second and 8 bits of quantisation per sample. In practice, rather than sending the code of the absolute value of each voice sample, the difference or delta between samples can be coded to lower the transmission bit rate. This form of analogue to digital conversion is known as *differential PCM* or delta communications [4.6]. There are many different types of binary coding, including non-return to zero and return to zero. Further details about these coding methods can be found in [4.7].

4.8 Keying techniques

Once a digital baseband is generated, it is necessary to transmit it. Like analogue, digital basebands occupy a similar band of frequencies and so it is vital to use modulation to obtain suitable transmission frequencies. For digital basebands, a further problem requires use of modulation techniques. A binary digital signal is effectively a switched direct voltage and this does not propagate efficiently along lines or through the atmosphere. Therefore, even if we don't want to translate the digital baseband in frequency, we must modulate a Voice Frequency (VF) tone for efficient transmission. The use of a digital baseband as a modulating signal is termed *keying*.

Digital keying techniques can be applied directly to the carrier. It is usual to apply them first to a VF tone and then to use the keyed tone as the modulating signal for a higher frequency carrier if required. We have seen that modulation techniques can be: Amplitude Modulation (AM), Frequency Modulation (FM) or Phase Modulation (PM). Keying techniques can be considered under these three headings in a similar manner, as we observed with analogue modulation.

A digital signal consists of near instantaneous shifts between carrier conditions and it is usual to describe keying techniques as shift keying of a particular carrier parameter. We have *Amplitude Shift Keying* (ASK), when a digital baseband is used to shift the amplitude of the carrier or VF tone, as well as *Frequency Shift Keying* (FSK) and *Phase Shift Keying* (PSK), a carrier's frequency and phase respectively.

4.8.1 Amplitude Shift Keying (ASK)

ASK is a digital keying technique that works by shifting the carrier's amplitude between two preset levels. ASK is commonly used with two amplitude levels. The most extreme form of two-level ASK is where the tone is switched on and off by the digital baseband, and is referred to as On/Off Keying (OOK) (figure 4.7).

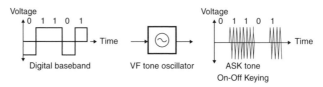

Figure 4.7: *Amplitude shift keying.*

ASK is effectively a DSB(FC) technique, as a digital baseband is directly multiplied by the VF tone; this doubles the bandwidth required for the digital signal when using ASK. (figure 4.8).

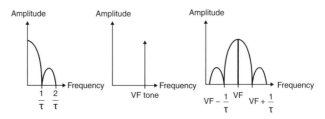

Figure 4.8: *Frequency spectrum of ASK transmission.*

If we consider a digital system capable of handling a single sine wave, we get a base bandwidth of $\frac{1}{2\tau}$.

ASK is used over short copper cables between a source of digital basebands and processing equipment. It is used as the keying technique for semiconductor lasers with optical fibres. ASK is not suitable for atmospheric propagation as the lack of signal during the transmission of a zero makes it vulnerable to noise. Further rapid keying of the signal on and off can cause large transmitter voltage transients leading to high power transmitter breakdown. There is little time for a system to settle down to a steady state frequency and so frequency variation transients are common. For atmospheric propagation, we need a signal that is always present.

4.8.2 Two-tone keying systems

To overcome the limits of ASK for atmospheric propagation or long distance transmission over copper cables, a two-tone keying system is used. One tone represents a 1 and the other represents a 0. A signal is always present, making a two-tone system less vulnerable to noise. Any recorded zeros are thus due to signal loss.

4.8.2.1 Frequency Shift Keying (FSK)

Frequency Shift Keying (FSK) relies on a Voltage Controlled Oscillator (VCO). A VCO is an oscillator that outputs a frequency dependent on the value of the externally applied voltage, in a similar manner to that discussed in frequency modulation. In figure 4.9 we see a VCO controlled by a digital baseband. If the baseband is a 1, the VCO outputs a frequency f_1, and if the baseband is a 0, the VCO outputs a second frequency f_2. This method gives constant amplitude signals with the tone frequency smoothly switching between frequencies f_1 and f_2.

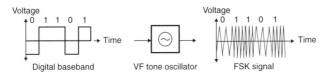

Figure 4.9: *Frequency shift keying.*

Each frequency f_1 and f_2 is considered as a single ASK signal with only one signal present at any one time. The frequency spectrum of an FSK signal is illustrated in figure 4.10. It appears as two ASK signals separated by a distance called the *shift*.

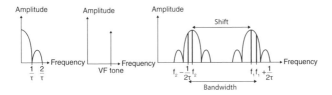

Figure 4.10: *Frequency spectrum of an FSK transmission.*

The FSK bandwidth requirement is larger than the equivalent ASK signal due to the shift. Actual bandwidth requirement runs from $f_1 + \dfrac{1}{2\tau}$ to $f_2 - \dfrac{1}{2\tau}$

$$\text{FSK bandwidth} = \left(f_1 + \frac{1}{2\tau}\right) - \left(f_2 - \frac{1}{2\tau}\right)$$

FSK bandwidth $= (f_1 - f_2) + \dfrac{1}{\tau}$

As $f_1 - f_2 =$ the shift, and $\dfrac{1}{\tau} =$ the baud speed

FSK bandwidth = the shift + baud speed   **(eq 4.9)**

Example 4.9: If $f_1 = 2.00$ GHz, $f_2 = 1.95$ GHz and $\tau = 1 \times 10^{-6}$ s, what is the FSK bandwidth (2 significant figures)?

Using: FSK bandwidth $= (f_1 - f_2) + \dfrac{1}{\tau}$

$$= (2.00 - 1.95) \times 10^9 + \frac{1}{1 \times 10^{-6}}$$
$$= 51 \times 10^6 \text{ Hz or 51 MHz}$$

FSK detection can be performed by two filters, one centred on f_1 and the other centred on f_2. The signal is applied to both filters and the detectors determine which frequency is present and reconstitutes the digital baseband. FSK has significant advantages over ASK, as not only is this a constant amplitude signal but also phase is continuously varying through the whole signal, so FSK is especially suitable for high power use. The main disadvantage of FSK is that it is produced by switching a single oscillator. At the switching speeds involved, the shift, and hence the capacity used, is limited. This is overcome using Frequency Exchange Keying (FEK) to provide larger shifts.

4.8.2.2 Frequency Exchange Keying (FEK)

FEK uses a two-tone signal but has two oscillators (figure 4.11).

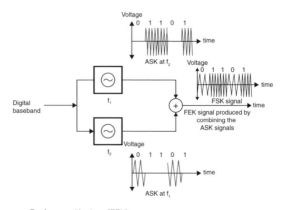

Figure 4.11: *Frequency Exchange Keying (FEK).*

The digital baseband is applied to both the oscillators to produce two independent ASK signals. This means that when f_1 is keyed on, tone f_2 is keyed off and vice versa. These two ASK signals are combined to give a two-tone FEK signal. The frequency spectrum of a FEK signal is the same as an FSK signal, because the only difference between them is in the *phase* of the frequency components and a frequency spectrum does not show phase.

FEK allows us to use wider shifts and higher baud speeds, but at the cost of losing phase continuity in the digital signal. The two oscillators will not be in phase at the switchover unless care is taken to match the two tones and the keying speed and is usually ignored in practice. Non-continuous FEK can only be used for low power signals where the transients produced are not disruptive. It is usual to use FEK for high speed, lower power applications and FSK for low speed, high power applications.

4.8.2.3 Phase Shift Keying (PSK)

Two-tone keying increases ASK bandwidth requirements by including a frequency shift. Another angle modulation system, *Phase Shift Keying* (PSK), provides a constant amplitude signal with no increase in bandwidth over ASK.

Simple PSK involves a phase reversal of 180 degrees, as shown in figure 4.12. This type of signal is sometimes called *Phase Reverse Keying* but is more commonly known as *BiPhase Shift Keying* (BPSK).

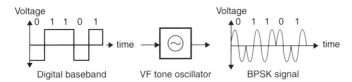

Figure 4.12: *BPSK transmission.*

The digital baseband shown has each transition from a + 1 to − 1 introduce a 180-degree phase shift in transmitted waveform. The BPSK spectrum is the same ASK but with the carrier or tone component missing. This is equivalent of DSB(SC) used in AM. The frequency of a BPSK signal is the same as that for an ASK signal (figure 4.8). However, BPSK has advantages over ASK as it is constant in amplitude with no carrier component transmitted.

4.9 Spread spectrum techniques

In telecommunications, spread spectrum techniques allow the signal to be transmitted to be spread over a much wider bandwidth than would normally be required from signal bandwidth alone. This allows us to exploit the Shannon-Hartley theorem (equation 4.4).

From the Shannon-Hartley law we see that the channel capacity, expressed as the maximum bit rate of a channel, depends on both bandwidth B *and* the S/N ratio.

$$\text{Capacity} = 2B \ log_2\left(1 + \frac{S}{N}\right) \qquad \textbf{(eq 4.4)}$$

For a given transmission, the capacity required by the channel is fixed by the data rate. As capacity depends on transmitted signal bandwidth and S/N, for a given capacity we can choose to use either a wide bandwidth and a low S/N ratio or vice versa. Using a very wide bandwidth transmission system means we sacrifice S/N ratio by reducing our transmitted power to a very low level and still maintain required capacity. This accounts for the considerable S/N advantages enjoyed by wideband systems such as digital signalling and FM.

Spread spectrum techniques can reduce transmission time by providing a higher capacity channel, allowing us to increase data rate. The spread spectrum techniques we will consider are pseudo-randomly varying the carrier frequency of a narrowband signal (*frequency hopping* or frequency agility), spreading the bandwidth to compress transmission time (*burst transmission*) and spreading bandwidth to create a low power density signal (*direct sequence spread spectrum*).

These techniques are made by multiplying the data $d(t)$ with a spreading signal $p(t)$ independent of the data. The channel will have an interfering signal $i(t)$. The original signal is recovered at the receiver using the same spreading code, synchronised with the received spread signal, to despread the signal. So the incoming signal $d(t) \times p(t) + i(t)$ is multiplied by p(t) to yield p(t) × d(t) × p(t) + p(t) × i(t) = $d(t) + p(t) \times i(t)$.

The first term is recovered using a filter of bandwidth 1/T, while the second term indicates the interfering signal has been spread over the spreading signal bandwidth. Even if the interfering signal is in the middle of the chosen band, only a tiny fraction of its power can pass through the filter. There are several applications of spread spectrum, including protection against interference, separation of users sharing a common medium, and reduction in energy density to meet frequency regulations, minimise detectability or provide privacy [4.8]. Spread spectrum has long been used by the military as an anti-jamming measure, but more recently its application has been extended to commercial communications [4.9–4.10].

Spread spectrum offers three major advantages over fixed frequency transmission, especially relevant to civilian systems:

1 Spread spectrum signals are highly resistant to narrowband interference. The process of recovering a spread signal spreads out the interfering signal, causing it to disappear into the background.

2 Spread spectrum transmission shares frequency bands with many types of conventional transmissions with minimal interference. Spread spectrum adds minimal noise to the narrow frequency communications and vice versa. As a result, bandwidth is used more efficiently.

3 Spread spectrum signals are hard to intercept. To a narrowband receiver, spread spectrum signals appear as a slight increase in background noise.

Further details about spread spectrum radio techniques can be found elsewhere [4.11].

4.9.1 Frequency hopping

This method effectively reduces transmission time. It also pseudo-randomly alters the transmitted signal carrier frequency during the message. Frequency hopping systems offer protection against disruption from other spectral users, but offers no protection against detection, which is a greater consideration for military users.

4.9.2 Burst transmission

Burst transmission is often used by military systems to complete the message transmission before an interceptor can register and jam the reception. A key difference is that burst transmission uses a very high data rate to compress the message transmission time. If a message takes 10 seconds to transmit at 60 bps, it will take 0.1 seconds at 6.0 kbps. Use of very high transmission rates can substantially reduce transmission time to prevent disruption. As digital signal bandwidth is equal numerically to half the data rate, burst transmission occupies a wide bandwidth so each signal occupies a large portion of available bandwidth; however, detection of such a wide bandwidth signal is comparatively easy.

4.9.3 Direct Sequence Spread Spectrum (DSSS)

In telecommunications, DSSS is a key modulation technique, and is probably the most sophisticated of spread spectrum techniques. DSSS uses a high speed code with a wider bandwidth than is actually needed by the message signal to spread the signal over a wide bandwidth and then remove the code (*despread*) from the signal in the receiver. So far, all the Modulo-2 addition has been between two signals having the same data rate. In DSSS, the spreading code has a much higher data rate (typically the order of M bps) compared with the signal (a few k bps). In DSSS, the message signal used to modulate a plain text bit sequence is a Pseudo Noise (PN) code, consisting of pulses of much shorter duration (and consequently larger bandwidth) than the message pulse duration itself. The message signal modulation has the effect of chopping up the message pulses and resulting in a signal that has a bandwidth nearly as large as the original PN sequence [4.13].

The string of PN code symbols are called *chips* each of which has a much shorter duration than an information bit. Each information bit is modulated by a larger sequence of much faster chips, so the chip rate is much higher than the information bit rate. Consider a signal spread with a spreading ratio of 5 (figure 4.13). Some of the uses of DSSS include Code Division Multiple Access (CDMA) and Wi-Fi networking.

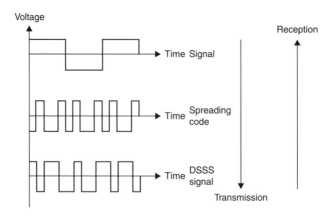

Figure 4.13: *DSSS process.*

We see from figure 4.13 that the DSSS signal ends up with a data rate equal to the spreading code and the signal is effectively spread over a much wider bandwidth.

The whole process is shown:

Signal	11111 00000 11111
Code	01001 01001 00101
Modulo-2 addition	------------------------------
Transmitted signal	10110 01001 11010

Received signal	10110 01001 11010
Code	01001 01001 00101
Modulo-2 addition	------------------------------
Recovered signal	11111 00000 11111

The despread signal contains the same power as the received signal, but as the bandwidth is compressed, power density is increased. This improves the signal S/N ratio, and also discriminates against interference as the signal, unless similarly coded, will not be despread. As we get a large power density improvement in the receiver, the DSSS signal can be transmitted with low power density – lower than the ambient noise, which makes a DSSS signal very hard to detect in the first place – and a low power consumption method. The resulting effect of enhancing signal to noise on the channel is called process gain.

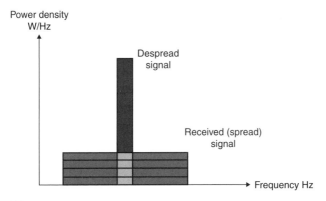

Figure 4.14: *DSSS sequence.*

The benefits of DSSS are the ability to share a single channel among multiple users, reduced signal to background noise, and resistance to jamming, There are many key satellite users of DSSS systems, such as satellite GPS.

A further benefit arises when several signals are spread using different codes. At the receiver, only the signal spread using the set code will be despread and this signal can easily be read, leaving the other signals as noise. Thus several different, independent users can use the same range of frequencies at the same time. This is the basis of a multiple access system called *Code Division Multiple Access* (CDMA), which gives every user the entire channel for all of the time and makes it almost impossible for other users to detect any one signal as their combination appears as a rise in the general noise level received by a scanning receiver.

CDMA is often compared with other multiple access techniques that have the same objective – that is, to share a communications media among several users. Time Division Multiple Access (TDMA) works by separating users in time using time slots, one per user. Frequency Division Multiple Access (FDMA) divides the frequency band into smaller bands, one per user. FDMA requires more complex implementation.

Different methods make different uses of frequency and time slots. Figure 4.15 shows the ways in which each transmission makes use of different proportions of frequency and time when in use.

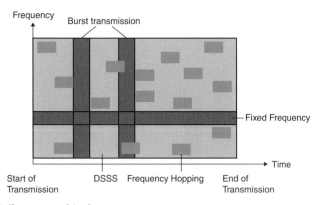

Figure 4.15: *Different uses of the frequency spectrum.*

A wide vertical spread shows wide use of the frequency spectrum and a wide horizontal spread shows a wide use of the time domain. Only frequency hopping and spread spectrum operate over a wide spread in each direction and only spread spectrum uses all the time and frequency available.

4.9.4 Civil Frequency Hopping Spread Spectrum (FHSS)

This is a method of transmitting radio signals by rapidly switching a carrier among many frequency channels, using a pseudo-random sequence available to both transmitter and receiver. Each available frequency band is divided into sub-frequency bands. Signals rapidly change or hop among these in a predetermined manner. Interference at a specific frequency only affects the signal during that short interval.

Civilian use of FHSS is commonly available in the 2.4 GHz band. Some walkie-talkies employ FHSS for unlicensed use on the 900 MHz band. Modern military radios use frequency hopping and employ separate encryption devices.

Bluetooth is a civilian wireless technology standard for exchanging data over a short distance, using short wavelength UHF radio waves in the 2.4–2.485 GHz band, from fixed and mobile devices [4.13]. In the future, smart radios and other new wireless devices will be able to avoid transmission congestion by switching almost instantly to adjacent unused nearby frequencies that they sense are clear of channel traffic [4.14].

4.10 Wireless technology examples

4.10.1 Civilian wireless technologies

The most common wireless technology uses radio. It can cover a range of fixed, mobile and portable applications, including two-way radio, cellular telephones and wireless networking services, which are essential to consumer and professional Marine VHF radio users.

Marine VHF radio covers the range between 156 and 162.025 MHz. Marine radio equipment is installed on all large ships and most seagoing craft. It is also used, with slightly different regulation, on rivers and lakes, for a wide variety of purposes including rescue services and communicating with harbours, locks, bridges and marinas, operating in the VHF band.

Standard handheld maritime VHF radio is mandatory on large seagoing vessels under Global Maritime Distress Safety System (GMDSS) rules. A marine VHF set is a combined transmitter and receiver and operates on standard international frequencies known as Channel 16 156.8 MHz, which is the international calling and distress channel. Transmission power levels vary between 1 and 25 W, giving a maximum range of up to 60 nm between aerials mounted on tall ships and hills, but as little as 5 nm between aerials mounted on small boats at sea level [4.15].

Modern marine VHF radios offer basic transmit and receive voice capabilities and Digital Selective Calling (DSC), which allows data distress signals to be broadcast with a single button press. Marine VHF uses half-duplex transmission, where communications take place one direction at a time. A transmit button on the set determines whether it operates as a transmitter or receiver. DSC equipment, part of GMDSS, provides all the functionality of voice-only equipment but can also automatically call a receiver equipped with DSC using a telephone-type number, known as a Maritime Mobile Service Identity (MMSI). DSC information is sent to the reserved Channel 70. When a receiver picks up the call, this active channel is automatically switched to the transmitter's channel and normal voice communication can proceed.

4.10.2 Military wireless technology examples

The armed forces make extensive use of communications systems, one example of which is the Bowman Military Communications Network, the latest tactical communications system used by the British armed forces, including the Royal Navy. The Bowman C4I system consists of a range of HF, VHF and UHF radio sets designed to provide secure integrated voice and data services to soldiers, vehicles and command HQs up to Divisional level [4.16].

Bowman replaced the ageing Clansman radio to provide a tactical voice and data communications system for joint operations across the British armed forces and is now fitted to over 15,000 military vehicles. The Royal Navy fleet and all the major helicopter types, including Merlin and Lynx, are fitted with Bowman equipment. Radios range between a 5 W man-pack up to a 50 W vehicle-mounted high-power radio, a smaller number of 100 W high-power, frequency hopping HF radios, and the UHF High Capacity Data Radio (HCDR) with a 225–450 MHz operating frequency range. Further details discussing the difficulties of the original Bowman procurement and implementation are discussed elsewhere [4.17]. Other armed forces around the world are currently developing similar systems.

4.10.3 Wireless links

Computers are often connected to networks using wireless links. The commonly used wireless methods are: terrestrial microwave, communications satellites, cellular systems, radio and spread spectrum technologies, and free space communications. Further details of these can be found elsewhere [4.18] but include:*Terrestrial microwave* LOS relay stations (50 km apart).

Communications satellites Satellites typically in geosynchronous orbit 35,400 km above the equator. Earth-orbiting systems receive and relay voice, data and TV signals.

Cellular systems Use radio communications technologies, dividing a region into multiple areas. Each area has a low-power transmitter or radio relay antenna to relay calls from one area to the next.

Radio and spread spectrum technologies Wireless Local Area Networks (LANs) use spread spectrum technology to communicate between multiple devices in a limited area. The IEEE 802.11 protocol defines wireless radio wave technology known as Wi-Fi.

Free space optical communication Uses visible or invisible light for LOS communications.

4.10.4 Li-Fi

Li-Fi (Light Fidelity) is a bidirectional, high speed and fully networked wireless communication technology similar to Wi-Fi, recently coined by Dr Harald Haas of Edinburgh University, UK. It is about 100 times faster than existing RF Wi-Fi implementations, reaching speeds of 224 gigabits (Gb) per second [4.19]. It is wireless and uses visible light or infrared communication instead of RF spectrum. It carries much more information, and is proposed as a solution to current RF bandwidth limitations.

Self-assessment questions

4.1   Describe the sequence of encrypting and decrypting an 8-bit binary signal using a PRNS, 10110110, illustrating your answer with an appropriate example.

4.2   Explain why a *minimum bandwidth* must be used when specifying digital communications channels, rather than an exact one.

4.3   If there are 256 levels per sample for speech, with 7000 samples per second calculate the specified bit rate and the minimum bandwidth for a communications channel using these figures.

4.4   Explain why digital signals have advantages over analogue ones.

4.5   Briefly explain, with a suitable diagram, how an analogue signal is converted into a PCM signal. Calculate the minimum channel bandwidth required to carry a PCM version of an analogue signal whose highest frequency is 30 kHz, choosing a suitable, practical sampling frequency. What will be the bit rate using 16 bits per sample?

4.6   Explain why increasing the bandwidth of a digital signal allows it to be successfully transmitted at a much lower power without loss of accuracy on reception.

4.7   Explain what is meant by frequency hopping and its advantages and disadvantages.

4.8   Discuss the problem of synchronisation and how it may be achieved.

4.9   Explain how a burst transmission is created and its advantages and disadvantages.

4.10   Explain how a DSSS signal is produced.

REFERENCES

[4.1] 'Certain Topics in Telegraph Transmission Theory', H Nyquist, Trans AIEE 47 (February 1928), pp. 617–644.

[4.2] www.write-numbers.com/en/bi/5200

[4.3] 'What is Encryption?', EFF Surveillance Self-Defense Project, n.d. Web. (22 April 2015). ssd.eff.org/en/module/what-encryption.

[4.4] en.wikipedia.org/wiki/Stream_cipher

[4.5] 'Public-Key Encryption in a Multi-user Setting: Security Proofs and Improvements', M Bellare (Springer, Berlin, Heidelberg, 2000, ISBN 978–3-540–67517–4), p. 1.

[4.6] *Digital Transmission Systems 3rd edition*, D R Smith (Springer Science and Business Media, 2003, ISBN 978–1-4020–7587–2).

[4.7] ibid. p 663.

[4.8] 'The Origins of Spread Spectrum Communications', RA Scholtz, IEEE Trans. on Comm, Vol COM-30, No.5 (May 1982), pp. 822–854.

[4.9] 'Spread Spectrum for Commercial Communications', DL Schilling and others, IEEE Comm. Mag. Vol. 29, No.5 (April 1991), pp. 66–79.

[4.10] *Spread Spectrum in Action*, J A Vincent, Radio Communication (August 1993) pt 1: pp.53–56, pt 2: pp.68–71.

[4.11] 'Spread-Spectrum Radio', DR Hughes and D Hendricks *Scientific American* (April 1998), pp. 82–84.

[4.12] *Communication systems* (4th edition), S Haykin (Wiley, 2008) p. 488. (retrieved 11 April 2015).

[4.13] en.wikipedia.org/wiki/Bluetooth.

[4.14] 'Cognitive Radio', S Ashley, *Scientific American* (March 2006) pp. 46–53.

[4.15] www.gov.uk/government/uploads/system/uploads/attachment_data/file/442648/MGN_
       324Corr.pdf

[4.16] en.wikipedia.org/wiki/Bowman_(communications_system)

[4.17] "LiFi internet: First real-world usage boasts speed 100 times faster than WiFi", A Cuthbertson
       (23 November 2015) (last accessed 10 February 2016).

[4.18] 'Ministry of Defence: Delivering digital tactical communications through the Bowman CIP
       Programme', House of Commons Committee of Public Accounts, Fourteenth Report of Session
       2006–07 www.publications.parliament.uk/pa/cm200607/cmselect/cmpubacc/358/358.pdf

[4.19] en.wikipedia.org/wiki/Computer_network

5

Radar

'[The scientist] believes passionately in facts, in measured facts. He believes there are no bad facts, that all facts are good facts, though they may be facts about bad things.'

Sir Robert Alexander Watson-Watt, speech to the
Empire Club of Canada (Jan 1948)

5.1 Early history of radar

For our discussion of maritime sensors, let us start with radar. The radar is the principal sensor that a Bridge Officer or Officer of the Watch (OOW) uses on any platform to ensure ship safety. Radar is an acronym for Radio Aid for Detection And Ranging and is a technique using radio waves to determine a target's location in space. Due to the speed of electromagnetic waves, detection and ranging happens virtually simultaneously. Clearly, optical systems cannot provide all the shipping information we require, especially from space, as cloud cover limits operational usefulness; in fact, worldwide cloud cover will vary between 40 and 80 per cent of the surface coverage. Hence, something that can 'see' through weather, and is available during the day as well as the night, is a clear navigational safety advantage. Marine radars are either X band (3 cm wave) or S band (10 cm wave) to provide bearing and distance of ship and land targets in the vicinity of the ship for collision avoidance and safe navigation at sea. Radars are rarely used alone or in isolation – commercial ships are integrated into a full system of marine instruments including sonar, chart plotters, radio, and emergency locators such as SART. In the busiest ports and harbours of the world, shore-based Vessel Traffic Services (VTS) equivalent to Air Traffic Control radar systems are used to monitor and regulate ship movements in busy waters. All ships are mandated to maintain a proper radar lookout, if it is available, to obtain early warning of risk of collision. Radar plotting or ARPA should be used to get the information of movement and the collision risk (bearing, distance and CPA) of other ships in the vicinity.

The story of modern radar begins during the rearmament of Nazi Germany in the 1930s. In 1935, the Scottish physicist Sir Robert Alexander Watson-Watt was approached as to whether radio waves might be used to detect aircraft approaching the English shores. His response to the Air Ministry, with the support of Sir Henry Tizard, resulted in a concentrated radar development programme. The problem of getting hold of a high power transmitter in such a short time was overcome by using a conveniently located high power BBC radio transmitter. Watson-Watt used an existing short wave radio receiver tuned to the 49 m wavelength of the BBC Empire Radio Service. The original radio receiver is shown in figure 5.1a, with corresponding valves in 5.1b (see plate section).

By March 1936, the first high masts, which could track aircraft at 80 miles, were built for the RAF at Bawdsey Research Station, Suffolk UK. It became the first link in the Chain Home (CH) system operated at, by modern terms, low frequencies (22–50 MHz), detecting propeller driven aircraft under good atmospheric conditions up to 150 km away, but much less during the passage of weather fronts. Nonetheless, radar dramatically increased British ability to detect targets beyond the visible horizon and provided some positional information as well. By September 1939, at the outbreak of World War II, 20 CH command stations were fully operational. Peak system power output was 350 kW, using dual transmitter towers and single reception antennas. The radar equipment used in the Battle of Britain was developed by a group of scientists initially based at the National Physical Laboratory in Teddington. They could detect reflected radio waves from moving bomber aircraft and, importantly, developed a system of practical operational procedures for using the information.

This network centric approach meant that assets of an at times significantly depleted RAF could be accurately vectored to intercept German bomber aircraft, acting as a significant 'force multiplier'. The critical role of the outcome of the Battle of Britain had an important knock-on effect: namely, by the autumn of 1940, key British radar experts and details of the cavity magnetron (operating at 10 cm with a 10 kW peak pulse) were sent to America – vital to American development of radar technology. Further details about radar's sensing role and its role in the Battle of Britain are provided in the references [5.1–5.3].

Research on airborne interception radar rapidly helped develop a range of special purpose radio sets. Radar not only enabled Allied bombers to provide a basic radar 'map' of land below the clouds at night but also provided a powerful aid in the hunt for surfaced U-boats. It forced U-boats to spend more time under water, reducing their mobility, and is considered crucial to the defeat of the U-boat menace in the Atlantic theatre of warfare. However, German U-boats also had a radar detection capability in their Naxos Radar Warning Aerial, which by 1944 enabled them to detect patrolling aircraft so that they would attempt to dive before being attacked.

The British lead in radar was consolidated after the war, particularly in radio astronomy. Bernard Lovell spent the war working on radar and particularly the H2S radar bombsight. After returning to research at The University of Manchester, and with the support of Nobel Laureate and BRNC-trained Patrick Blackett, Lovell's group became world leaders in studying meteors and detecting signals from radio sources in Andromeda, a separate galaxy. Lovell's Jodrell Bank radio telescope was the only instrument in the West able to track Sputnik and other Soviet space probes, raising radio astronomy – and Manchester – to a prominent position.

5.2 The echo ranging principle

Most radars, but not all, operate using the *echo principle*. If electromagnetic wave pulses strike an object with the 'right' aspect, wave energy is reflected directly to the radar.

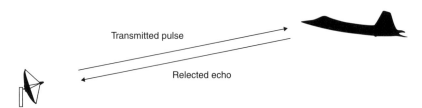

Transmitted pulse

Relected echo

Figure 5.2: *Echo ranging principle.*

If the time (t) from the transmitted pulse to the received echo is measured, knowledge of the wave speed allows us to calculate contact range accurately. Since electromagnetic waves all travel at the speed of light, the distance they travel out and back to the aircraft in figure 5.2 is given by: distance = c t, where t is the elapsed time, while the contact range (R), which is half the total distance travelled, is given by:

$$R = \frac{ct}{2} \qquad\qquad \textbf{(eq 5.1)}$$

Since 'c' does not vary significantly, time is proportional to contact range. This relationship is used in pulsed radars to measure contact range and is known as *pulse delay ranging*. This principle is used for all types of active electromagnetic systems such as Light Detection And Ranging (LIDAR) and is applied to sonar, which uses sound energy, travelling at more modest speeds.

Example 5.1: A radar transmits a pulse of microwave energy which takes 3 ms to return an echo to the radar. What is the contact range?

Using the equation $R = \dfrac{ct}{2}$, $R = (3 \times 10^8 \times 3 \times 10^3)/2 = 4.5 \times 10^5$ m or 450 km.

5.3 Fundamental radar parameter definitions

The application of a radar determines the chosen parameters of the radar. By appropriate selection of the radar parameters, it can be *optimised* for its intended use. We define a pulsed radar as one that transmits regular pulses over long periods of time.

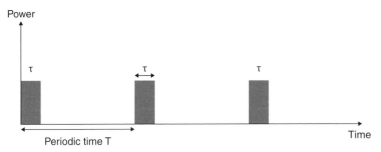

Figure 5.3: *Transmission of regular radar pulses.*

Key parameters are: the rate pulses are transmitted, the number of pulses per second, and its pulse duration (τ), which is short for high definition navigation radar (0.08 microseconds) but up to 2–3 microseconds for long range search radar. The rate pulses are transmitted is the *Pulse Repetition Frequency* (PRF) and is the time between regularly spaced pulses, marked T in figure 5.3, referred to as the *Pulse Repetition Interval* (PRI). The relationship between the PRI and the PRF is analogous to the relationship between radar frequency and wave period.

Basically: $T = PRI = \dfrac{1}{PRF}$ **(eq 5.2)**

Note: PRF and radar frequency are not the same.

Monitoring the radar PRF provides information about the *type* of radar operating, e.g. the PRF of maritime surveillance radar is typically 700–800 pulses per second (pps).

Example 5.2: A radar transmits pulses at 10 GHz every 1 ms. What is the radar's PRF (1 significant figure)?

Using the formula:

$$PRF = \frac{1}{T} = \frac{1}{1 \times 10^{-3}} = 1000 \text{ pulses per second.}$$

Transmitted radiation frequency, beam width and Antenna Rotation Rate (ARR) are important parameters for radar design and identifying the transmitting radar.

Peak power (P_p) is the peak power transmitted during each pulse. Figure 5.4 shows the radar transmitting for a small period each cycle, so average power (P_{AV}) is a better guide to power transmitted by a radar set. Average power is given by:

$$P_{AV} = P_p \times \tau \times PRF = P_p \times \frac{\tau}{T} \qquad \textbf{(eq 5.3)}$$

$\frac{\tau}{T}$ is the *duty factor* and shows the fraction of time the radar transmits.

Example 5.3: What is the average power level if peak power is 1 MW, and $\frac{\tau}{T} = 0.1$ (1 significant figure)?

Since : $P_{AV} = P_p \times \frac{\tau}{T}$

$P_{AV} = 1 \text{ MW} \times 0.1 = 0.1 \text{ MW}$

Equation 5.3 and figure 5.4 indicate three methods of increasing the average transmitted power from its initial value exist, namely:

(i) increase PRF (decrease T)–figure 5.4a

(ii) increase pulse duration (τ)–figure 5.4b

(iii) increase peak power (P_p)–figure 5.4c.

Digital Sequence *Sara-Kate Lavers © 2016*

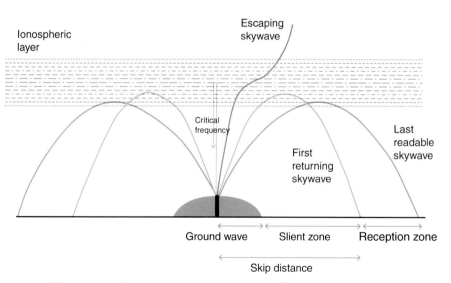

Figure 2.5 *The refractive effect of the ionosphere on waves at different initial angles of incidence to a layer, showing the different paths of waves totally internally reflected, and those that pass through the layer.*

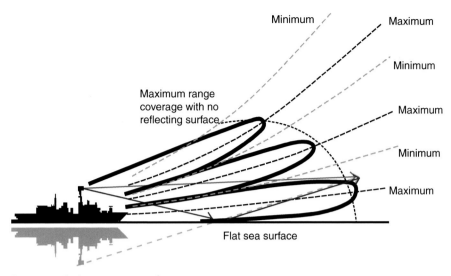

Figure 2.14 *Radar coverage interference.*

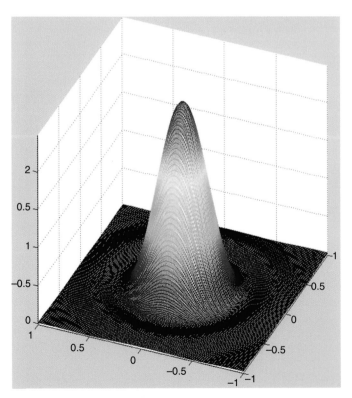

Figure 3.7 *Cylindrical spreading of Bessel function.*

Figure 5.1a *Original CH radio receiver.* Figure 5.1b *CH valve.*

Figure 5.17 *Cutaway inside of a magnetron, showing the cavities where radio frequency waves are induced.*

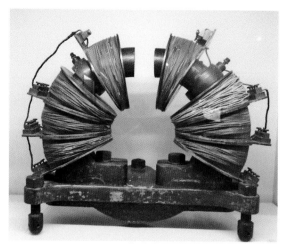

Figure 5.18 *The original electromagnet in the first magnetron experiments.*

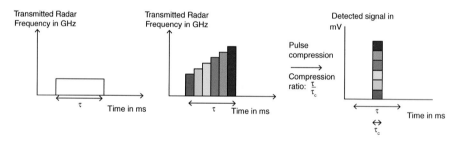

Figure 6.9 *(left) Conventional pulse, (middle) FM pulse, (right) receiver compressed pulse.*

Figure 6.17 *Seaspray 5000E radar. (Courtesy of Leonardo Airborne and Space Systems Division.)*

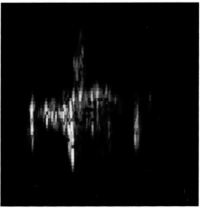

Figure 6.19 *Oil rig (left), ISAR radar image of the oil rig (right). (Courtesy of Leonardo Airborne and Space Systems Division.)*

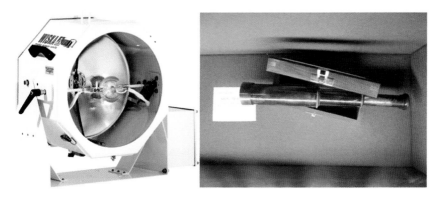

Figure 7.1a *SKS 57 Suez Canal searchlight. (Courtesy of WISKA.)* Figure 7.1b: *Naval telescope.*

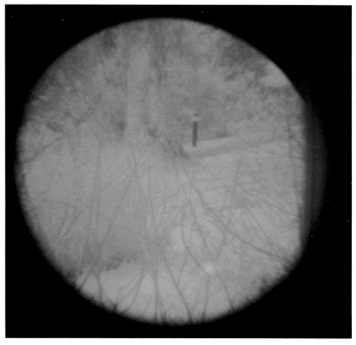

Figure 7.7 *Typical green phosphor II 'night vision' image.*

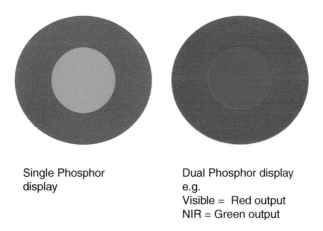

Single Phosphor
display

Dual Phosphor display
e.g.
Visible = Red output
NIR = Green output

Same Overall Radiated Intensity but different spectra

Figure 7.9 *The difference in contrast going from a single phosphor display output to a dual phosphor display output is clear.*

Figure 7.10 *Greyscale, true colour, dual colour system.*

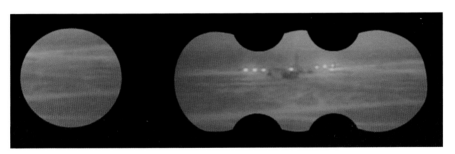

Figure 7.11 *(left) Typical NGV FOV. (right) Wider FOV system under test.*

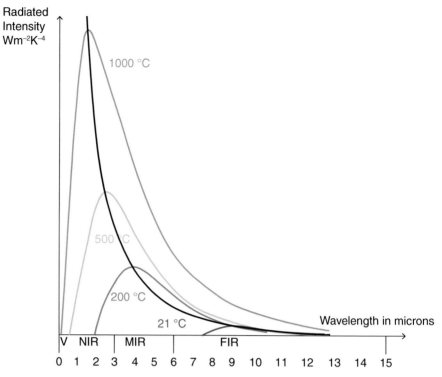

Figure 7.14 *Radiated power per unit surface area emitted as a function of wavelength.*

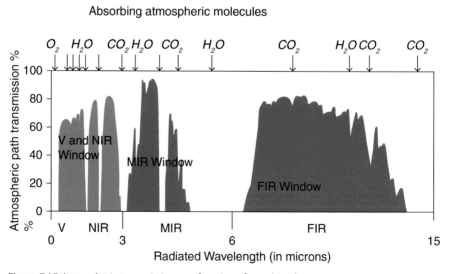

Figure 7.15 *Atmospheric transmission as a function of wavelength.*

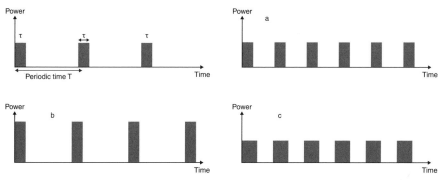

Figure 5.4: *Increase of PRF (a), peak power (b) and pulse duration (c).*

However, detection at a given range depends upon several factors.

(a) Increase PRF (decrease τ).

Increasing PRF will decrease the time interval between pulses.

(b) Increase pulse duration (τ).

Increasing pulse duration allows longer range detection as the pulse has more energy. However, increasing duration means short range returning echoes are not processed; so it is inappropriate for short range navigation to have a long duration pulse. The longer the pulse duration, the less time there is for cooling the radar, since more power produces more heat than it emits microwaves.

(c) Increase of peak power (P_p).

Increasing peak power increases range, but it also increases the likelihood of detection, giving a radar range advantage to anyone listening at these frequencies. This range advantage arises because of the *double* geometric spreading out and back, and twice the round trip atmospheric losses, incurred against one-way detection, with only one geometric spreading and one propagation path's scattering and absorption losses. Thus, the greater the power transmitted, the greater the range advantage possible. Although not a problem for civilian radar, this is a major issue for military maritime surveillance.

5.4 Maximum Detection Range (MDR)

The equation for radar MDR came from the requirement to predict likely detection ranges in a clear atmosphere. MDR depends on transmitted power, target reflecting properties, antenna size and receiver sensitivity. The MDR is the maximum range a radar can detect a chosen target based on several key parameters, a combination of controllable factors: P, G, tot, Aeff, Smin, and uncontrollable factors: target Radar Cross Section (RCS), environmental and weather effects. The equation is represented by:

$$\text{MDR} = \sqrt[4]{\frac{P_{AV} G \sigma t_{ot} A_{eff}}{(4\pi)^2 S_{min}}} \qquad \textbf{(eq 5.4)}$$

where G = the antenna gain,

σ = the target RCS (a measure of the contact echo size seen by the radar),

t_{ot} = the time a radar contact is illuminated by a radar beam,

A_{eff} = the effective receive antenna size,

S_{min} = the minimum signal energy required by the receiver, and

P_{AV} = average transmitted power.

Any change in these alters the MDR.

Wave losses increase with increasing frequency and the MDR falls. Quality of operator training, essential yet hard to quantify, is also important.

Example 5.4: If the RCS is decreased by a factor of 16, by how much will the radar MDR be reduced from its original value?

Using the equation: $\text{MDR} = \sqrt[4]{\frac{P_{AV} G \sigma t_{ot} A_{eff}}{(4\pi)^2 S_{min}}}$,

and considering the relative changes of RCS, then:

$$\frac{MDR_{new}}{MDR_{old}} = \sqrt[4]{\frac{\sigma_{new}}{\sigma_{old}}}$$

So:

$$MDR_{new} = MDR_{old} \times \sqrt[4]{\frac{\sigma_{new}}{\sigma_{old}}}$$

$$MDR_{new} = MDR_{old} \times \sqrt[4]{\frac{1}{16} \times \frac{\sigma_{old}}{\sigma_{old}}}$$

Thus:

$$MDR_{new} = \frac{1}{2} \times MDR_{old}$$

Hence, to make significant MDR reduction requires a really large RCS reduction!

Example 5.5: If RCS is decreased by a factor of 50, and the radar gain (G) is increased by a factor of 10, by what amount must the radar average power be increased to observe *no difference* in theoretically predicted MDR?

Using $MDR = \sqrt[4]{\frac{P_{AV}\, G\sigma t_{ot}\, A_{eff}}{(4\pi)^2 S_{min}}}$, changes to the gain and average power level P_{AV} must 'equal' the

changes to the RCS for the MDR to remain unchanged: ie $(P_{AV}\, G\sigma)_{new} = (P_{AV}\, G\sigma)_{old}$

Hence: $(xP_{AV}10G\frac{\sigma}{50})_{new} = (P_{AV}\, G\sigma)_{old} = (xP_{AV}\, G\frac{\sigma}{5})$

Thus: $x = 5$

Radar must increase output fivefold, a difficult requirement to achieve in practice without use of phased array technology.

Note: Increase of all the parameters on the top of the equation will increase MDR but increasing sensitivity (increasing S_{min} required) reduces MDR.

5.5 Maximum Unambiguous Range (MUR)

Correct range measurement depends on correctly recorded echo time. Timing errors will result in incorrectly recorded echo range. With pulse delay ranging, range to a ship contact is determined by measuring elapsed time between transmitting a pulse and receiving an echo.

Recalling $R = \frac{ct}{2}$ **(eq 5.1)**

It is vital that all received echoes are assigned to the correct transmitted pulse to calculate the correct range. If this doesn't happen, calculated range will be *ambiguous* or unclear. All indicated ranges are true ranges until a contact echo arrives back just as the next pulse is transmitted.

This problem is illustrated in figure 5.5.

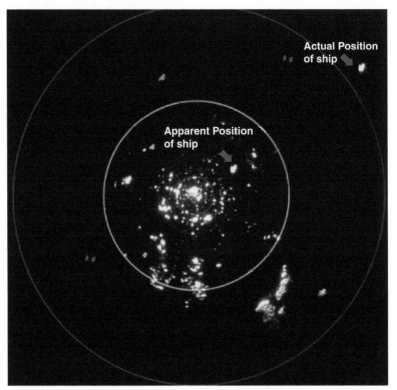

Figure 5.5: *Ambiguous echo is made apparent by doubling T (halving PRF).*

The indicated echo appears closer than it is because the echo returned after the next pulse has gone out. To counter this, a longer interval of time should be used so the most distant echoes are returned before the next pulse is transmitted.

Ambiguous echoes are detected by routinely altering PRF in navigation radars. True indicated ranges don't alter but apparent ranges will. Some radars carry out this process automatically. The PRF sets the MUR and the MUR is found by considering speed = distance travelled / time taken (figure 5.6).

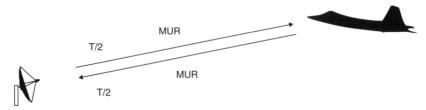

Figure 5.6: *Calculation of MUR.*

From the earlier range equation $R = \dfrac{ct}{2}$ **(eq 5.1)**, for the out MUR and back MUR over the total elapsed time T, we substitute this into eq 5.1 to get:

$2MUR = cT$

And rearranging:

$$MUR = \frac{cT}{2} \qquad \textbf{(eq 5.5)}$$

Given $T = PRI$ and $PRI = \dfrac{1}{PRF}$, we rewrite this, yielding:

$$MUR = \frac{c}{2PRF} \qquad \textbf{(eq 5.6)}$$

Example: 5.6: If the PRF = 1000 s⁻¹, what is the MUR (2 significant figures)?

Using the equation: $MUR = \dfrac{c}{2PRF}$ and substituting:

$$MUR = \frac{3 \times 10^8}{2 \times 2000} = 75 \ km$$

5.6 Data rate

A major parameter affecting reflected echo size for rotating conventional radar is the power striking a target. The incident power is given by the number of pulses striking the contact multiplied by the power in every pulse.

The number of pulses striking a contact is given by the time the contact is illuminated by radar multiplied by the PRF:

$N = t_{ot} \times PRF$ and,

since time on target $t_{ot} = \dfrac{a_H \times 60}{360 \times ARR} = \dfrac{a_H}{6 \times ARR}$ $\qquad$ **(eq 5.7)**

N is the number of pulses striking a target and t_{ot} is the target illumination time, giving:

$$N = \frac{a_H \times PRF}{6 \times ARR} \qquad \textbf{(eq 5.8)}$$

Example 5.7: If the beam width is 1 degree, PRF = 2000 s^{-1} and ARR = 60 RPM, what is N, the number of pulses striking a target (1 decimal place)?
Using the equation:

$$N = \frac{a_H \times PRF}{6 \times ARR}$$

And substituting:

$$N = \frac{1 \times 2000}{6 \times 60} = 5.6$$

At least five pulses strike the target every revolution.

5.7 Angular resolution

A radar's *angular resolution* is its ability to separate different contacts at similar range, but on slightly different bearings. A radar can resolve two contacts in bearing provided they have angular separation more than the horizontal beam width, i.e. a radar shouldn't receive echoes from both contacts simultaneously (figure 5.7).

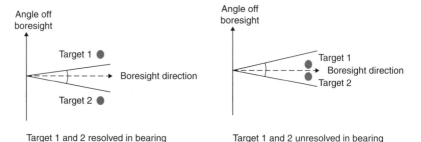

Figure 5.7: *Angular resolution.*

Pointing radars measure large inclination and have both horizontal and vertical angular resolutions (the horizontal a_H and vertical a_V beam widths respectively), given by slight modification of the generic beam width equation:

$$a_H = \frac{60\lambda}{D} \text{ and } a_V = \frac{60\lambda}{D}$$

Air search radar vertical beam width must be typically large, 65–70 degrees, while horizontal beam width will be typically 2–3 degrees (figure 5.8). Although 'ideal' to have a horizontal search beam width 0.5–1.0 degrees, practical wavelength limitations and limitations on radar size (if too large, it affects ship stability).

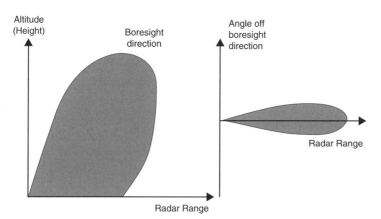

Figure 5.8: *Search radar coverage (VCD) (left) and Horizontal Coverage Diagram (HCD) (right).*

Typical navigation radar use is primarily for ship safety and collision avoidance, but secondary functions are needed, e.g. manoeuvring, blind pilotage, aid to navigation, small craft and helicopter control. For aerial heights of 10 m and a 2–3 m antenna, a radar can detect high land masses at over 50 km range under normal weather conditions.

Satellite and fire control radars are called *pointing radars* because they are very narrow (under 1 degree) and track in bearing and elevation. Generally, pointing radars do not rotate, but track using a variety of tracking techniques. Typical coverage diagrams for pointing radars are shown (figure 5.9).

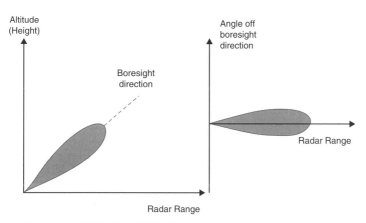

Figure 5.9: *Pointing radar VCD (left) and HCD (Range).*

5.8 Range resolution

Consider a pulse leaving a radar transmitter, as illustrated (figure 5.10). Waves occupy a certain length in space; the distance occupied by each pulse is velocity multiplying pulse duration = cτ metres.

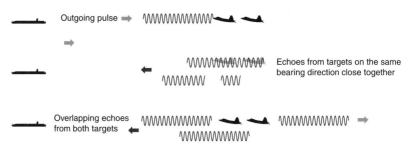

Figure 5.10: *Range resolution.*

If a pulse strikes contact A, some energy is reflected but some continues on to strike contact B. For two contacts to be distinguished, they must produce two separate echoes, i.e. the pulses mustn't overlap on return to the radar or they are interpreted as one echo.

The transition from overlap to no overlap is when the separation or range resolution $R = \dfrac{ct}{2}$; it is called the *range resolution* and is given the symbol RR.

$$\therefore \text{RR} = \frac{ct}{2}$$

Example 5.8: A navigation radar pulse is emitted with a duration of 0.1 microseconds. What is the range resolution achieved and its minimum detection range (2 significant figures)?

Using $R = \dfrac{ct}{2} = \dfrac{3 \times 10^8 \times 0.1 \times 10^{-6}}{2} = 15 \text{ m}$

A radar cannot resolve range below the range resolution. The smallest range step measured is the range resolution. As the radar receiver is switched off during transmission, range resolution is also a radar's *minimum range* as it isn't processing returning echoes during this time. Examples of range resolution for different pulse durations are shown in Table 5.1.

Pulse duration (τ)μs	Range resolution (cτ/2) m
50	7500
5	750
0.5	75

Table 5.1: *Pulse duration vs range resolution.*

5.9 Primary and secondary radars

Radars can be divided into primary and secondary radars. A primary radar requires no co-operation from a vessel under scrutiny and obtains range and bearing information with the echo principle. Secondary radars require vessel co-operation to provide additional information. The most common military application of secondary radars is Identification Friend or Foe (IFF). A radar transmitter, the interrogator, transmits a coded pulse at the target aircraft or ship. When the vessel receives this pulse a transponder is triggered and transmits its coded identification pulse back to the interrogator. This pulse is received by the interrogating unit, decoded and identification obtained. The latest generation of military IFF is the Mode S IFF, which provides multiple information formats to a selective interrogation. Each aircraft is assigned a fixed 24-bit address (figure 5.11).

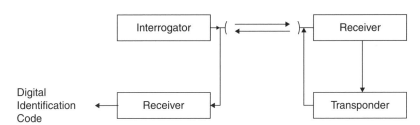

Figure 5.11: *IFF system.*

Where there are multiple targets it is very important that the IFF response is correctly associated with its contact. It is usual for an interrogator antenna to be mounted on a surveillance radar's antenna. Some secondary radar systems use the pulses of a primary radar as the interrogator, e.g. a system for identifying helicopters uses a ship's navigation radar as the interrogator.

5.10 Typical radar characteristics

5.10.1 Surveillance radars

Surveillance radars have swept, fan-shaped beams, usually produced by conventional rotating antenna, but may be produced by a controllable phased array. The beam may be swept over 360 degrees, for all round coverage, or across a restricted angle (sectored search), but the theory remains the same. Most of the performance characteristics are applicable to surveillance radars, but the actual values of the parameters chosen will depend upon the precise radar purpose. Table 5.2 shows the four main classes of surveillance radar, their desired characteristics and bias of the parameters.

Surveillance type	Characteristics	Parameters
1. Short range high definition (navigation).	Good target discrimination. High resolution in range and bearing. Small targets detected.	High frequency. High PRF. Very short τ. Fan-shaped beam with a vertical beam width of 30° to allow for a 15° ship roll.
2. Long range air surveillance.	Long MDR. All round coverage.	Low frequency. Low PRF for long MUR. Peak power limited so long τ for high average power. Poor range resolution – improved by pulse compression. Wide horizontal beam width due to low frequency and antenna size restrictions.
3. Medium range surface search.	Good surface and low air coverage. Good range and angular resolution. Good target discrimination.	High frequency. Medium PRF. Short τ. Fan-shaped beam with a vertical beam width of 30° to allow for a 15° ship roll.
4. Medium range air surveillance.	Good air coverage. Good range and angular resolution. Good target discrimination.	High frequency. Medium PRF. Short τ. Large vertical fan-shaped beam for air detection.

Table 5.2: *Typical characteristics of different surveillance radar.*

5.10.2 Pointing radars

Civilian small dish radars are useful for satellite communications links. The main military use of pointing radars are tracking and guidance functions associated with fire control. These radars are usually mounted together with the guidance system of the missile they control. Since they point at a target, these radars have a narrow *pencil beam*. Pointing radars commonly use a parabolic dish to produce this pencil beam with high PRFs to obtain a high data rate and PRF stagger or switching to resolve range ambiguities.

5.11 Doppler radars

Velocity information can be obtained by two common methods: either the analysis of a target's track indirectly using Track While Scan (TWS) or directly with the Doppler effect, which can act as a filter to remove unwanted real stationary contact returns.

5.11.1 The Doppler effect

When there is relative motion between a ship's radar and another vessel, received frequency differs from that emitted. If range is closing, received frequency will be higher than that transmitted, and if range is opening, received frequency will be lower than that transmitted.

The difference between the received and transmitted frequencies is called the *Doppler shift* and was first investigated by Austrian physicist Christian Andreas Doppler (1803–1853). In 1842 he observed that the frequency of a wave depended on the *relative speed* between the source and the observer. The Doppler shift is given by:

$$Doppler\ shift = f_{Received} - f_{Transmitted} \qquad \textbf{(eq 5.9)}$$

In target detection, relative motion between transmitter and target along the direct line between them yields a Doppler shift at the target and another equal Doppler shift on echo reception.

For all echo problems, the Doppler shift is given approximately by:

$$Doppler\ shift = 2f_{Transmitted} \times \frac{Relative\ Velocity}{c} \qquad \textbf{(eq 5.10)}$$

where $f_{Transmitted}$ is the transmitted frequency and c is the speed of microwave radiation. Doppler shift measurement gives direct information about the relative velocity between a transmitter and a target on the direct line between them.

Example 5.9: If the relative velocity between a ship's radar and another vessel is 10ms⁻¹, for a radar at a transmitted frequency of 10 GHz, what is the resulting Doppler shift (1 decimal place)?
Using the equation:

$$Doppler\ shift = 2f_{Transmitted} \times \frac{Relative\ Velocity}{c}$$

$$Doppler\ shift = 2 \times 10^8 \times \frac{10}{3 \times 10^8} = 6.7\ mps$$

5.11.2 Use of Doppler radar for target identification

If there is relative motion between a radiation source and a target, a Doppler shift in frequency occurs, which helps measure a target's relative speed towards a radar (figure 5.12).

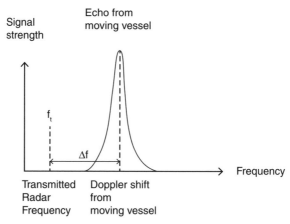

Figure 5.12: *The body of relatively moving platform generates a narrow spread of frequencies shifted to either a higher or lower frequency from that transmitted.*

Some targets contain moving parts that reflect microwaves (e.g. a fixed wing propeller driven aircraft or helicopter with rotor blades). Moving parts return a Doppler-shifted echo characteristic of the relative velocity of that part with regard to the radar providing a Doppler spectrum. Spectrum analysis provides specific information about unknown radar returns (e.g. helicopter rotor blade speed), which, if compared with a computer library of targets, can allow identification. Doppler radar has been used for military and civilian weather forecasting since early 1990, as its unique capabilities and suite of automated algorithms allow forecasters to accurately track precipitation intensity as well as detecting severe weather signatures and wind shear. Technology has become more advanced with use of combined pulse-Doppler radars; the motion and unique characteristics of rain droplets are now detected, allowing forecasters to make more accurate assessments of current and future weather [5.4–5.5].

5.12 Radar antennas

5.12.1 Dish antennas

The simplest and most common radar antenna is, however, the dish or *parabolic reflector*, which acts as a focusing mirror for radiation and is similar to a torch's focusing reflector (figure 5.13). If a wave source is placed at the focal point of a parabolic dish, it produces a highly directional beam.

Waveguide feeds are used in the source and receiver for parabolic reflectors. However, this leads to problems in supporting the weight of the feed. This is overcome by placing a sub-reflector in front of the focus and feeding it through a hole in the main reflector. The sub-reflector is supported on thin metal struts. A common shape for the sub-reflector is a hyperbola and the resulting antenna is the Cassegrain antenna illustrated in figure 5.13a. Cassegrain and Gregorian (5.13b) antennas are used in civilian satellite communications and military tracking radars.

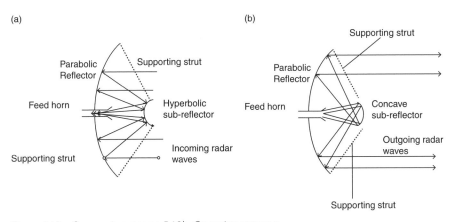

Figure 5.13a: *Cassegrain antenna.* 5.13b: *Gregorian antenna.*

Cassegrain antennas have a hyperbolic sub-reflector (convex) type but it is possible to use a *concave* reflector (Gregorian antenna).

A parabolic dish antenna gives conical or pencil beams, which are narrow in the vertical and horizontal directions. Beam width depends on the dish diameter and is given by:

$$a = 60 \times \frac{\lambda}{D}$$ (**eq 5.13**) where D is the dish diameter.

At high frequencies, dish antenna of reasonable size can achieve very narrow beam width. As we reduce transmitted frequency (increasing wavelength), we need increasingly larger dishes to maintain the same beam width. Below 5 GHz, dish diameter is too big for mobile use and mobile antennas are restricted to the microwave region.

> **Example 5.11:** A radar dish is used to produce a narrow beam. The radar frequency used is 20 GHz and the dish is 3 m in diameter. What is the beam width produced (1 decimal place)?
>
> $$a = 60 \times \frac{\lambda}{D} = \frac{60c}{fD} = \frac{60 \times 3 \times 10^8}{20 \times 10^9 \times 3} = 0.3 \, m$$

Until the advent of phased array radar, it was important but difficult to find a vessel or target's position accurately. As well as range, this involves obtaining the bearing of a surface target and for an air target bearing and height. Most air warning radars didn't actually obtain both bearing and height as there were few efficient 3-D radars. The height of a target was usually found with use of dedicated height finding radars or tracking/fire control radars known as pointing radars. To find bearing or target height accurately, the radar had to have a narrow beam width in the relevant direction – horizontal for bearing, vertical for height. It is important for early warning radar to detect aircraft approaching at any height, so these generally had narrow horizontal beam width and a broad vertical beam width. All round coverage was obtained by rotating the antenna, although phased arrays offer a better alternative solution, introduced next.

5.13 Phased arrays

Phased array radars are growing in use and considered as a single fixed aperture but are actually a large array of equally spaced sources covering the area of a single aperture, which for real vessel detection appear as a single focused beam. One advantage of arrays is that they are steered electronically and don't require mechanical beam steering, although some radar may do both. Arrays are composed of a number of sub-arrays, each of many hundreds of either phase shifting elements (passive array) or transmitting microwave emitting elements (active array) (figure 5.14).

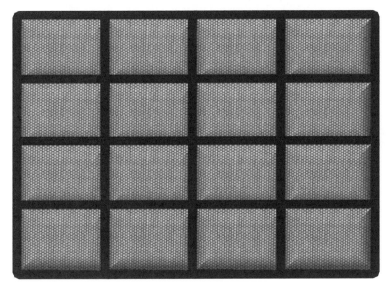

Figure 5.14: *Phased array (16 sub-arrays of many hundreds of elements).*

Phased arrays will be considered further in chapter 6.

5.14 Radomes

Most radar antennas are vulnerable to the two most serious effects of weather: wind and precipitation. As well as physical damage, precipitation causes changes in the antenna's electrical properties. Low frequency antennas are less affected by precipitation (especially ice build-up) and, being often of open mesh construction, are less vulnerable to wind. To protect radar antennas, they may be enclosed in radomes (figure 5.15), plastic structures designed for strength, yet transparent at the transmission frequency. In practice, some radiation is absorbed and scattered by the radome. Increased use of Frequency Selective Surfaces (FSS) in military radar means that only the transmitting radar's signal will pass both in and out of the radome unimpeded. Over time, water trapped in GRP can degrade the radome integrity, due to a microwave 'freeze–thaw' action.

Figure 5.15: *Typical radome. (Image courtesy of Kelvin Hughes.)*

5.15 The radar block diagram

A radar transmitter produces regular, high-powered pulses of radiation. Microwave power source in early sets was switched on and off by high voltage pulses produced by a pulse-forming network. Unfortunately, output frequency wasn't stabilised at a steady fixed frequency, due to transient waveform generation. A consequence of unstable frequency is an apparent Doppler shift, which may be due to instabilities in radar transmitter rather than actual target motion. The power source produced the electromagnetic pulse and provided all the system amplification, which resulted in wasted power through the entire radar system. Instead, the modern Master Oscillator Power Amplifier (MOPA) structure (figure 5.16) has a master oscillator, providing a continuous fixed frequency and low amplitude wave with both reference pulses to the receiver and ensuring that 'sculpture and shaping' of the pulse is achieved first at low power levels.

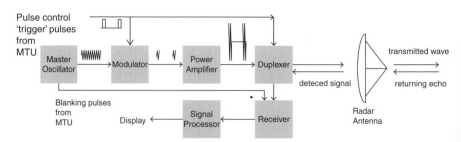

Figure 5.16: *Radar Master Oscillator Power Amplifier structure.*

The master oscillator is a continuously running, low powered oscillator, with greater stability than pulsed, high power oscillators. It also acts as a stable reference frequency for the receiver and as an input to the detection stage.

Modulator details include generation of a 0.08 microsecond navigation pulse. Logic trigger level pulses are created in the timing generator. The rapid decay from several hundred volts to 0 V will generate a high power pulse to enter the Pulse Forming Network (PFN), a bank of inductors and capacitors producing a high voltage 2-microsecond pulse.

Transmitter and receiver share a common antenna (normal configuration), with a device called a *duplexer* that guides all of the transmitted pulse energy to the antenna and prevents any energy leaking into the receiver, known as *receiver blanking*. Note: the radar set cannot receive and process echoes while transmitting.

To provide pulses at required intervals, a clock is needed. The device that does this is the 'trigger unit', which provides trigger pulses from a PFN to control the modulator and synchronise displays. Modulation of the continuous wave output with a trigger pulse produces the modulator output. All radars have an internal trigger unit, but it is usual to co-ordinate firing of a ship's radars with a Master Trigger Unit (MTU), providing trigger pulses to the ship's radars and receiver blanking pulses to non-firing radars, such as Satellite Communications (Satcom), which may be inadvertently damaged by high powered radar pulses. After amplification by the power amplifier, power is transmitted at an appropriate high power level.

5.15.1 Comment on radar RF microwave production

When radar was first invented, it was very hard to generate sufficient radio wave frequencies at 3 GHz until the invention of the magnetron. This device allowed the much easier creation of 3 GHz waves at high power. It typically consisted of a solid copper block with slots or circular chambers machined into it. In the centre of the block, a cathode was heated by a small voltage (figure 5.17, see plate section).

The magnetron was a new radar valve generating a high level of microwave power at a suitable high frequency. It transformed radar for submarine hunting, night fighter patrols, anti-aircraft gunnery and bomb aiming. The magnetron was developed in 1940 by John Randall and Harry Boot at Birmingham University and it was quickly shared with their American allies. They required a high power electromagnet to provide the strong magnetic field the magnetron needed, as shown (figure 5.18, see plate section). They attributed their success on 21 February

1940 to the one day on which all their items of laboratory hardware happened to work at once!

The copper body has a 2 kV pulse applied. If the cavities are held in a vacuum evacuated glass bulb, the difference in potential difference causes a stream of electrons to move towards the positive anode (body) from the negative cathode. If a strong magnetic field surrounds the copper body, electrons leaving the cathode follow curved paths, passing the slot entrances. As these electrons pass each cavity, an electrical wave sweeps into the cavity, setting up oscillatory standing waves, whose frequency is determined by the cavity's size. Radio waves are removed from the magnetron via a short metal dipole inserted into one of the machine cylinders. These radio waves are then led to the antenna through a waveguide.

Modern day magnetron radar, such as the Kelvin Hughes 10 kW X-band radar, are available with a range of antenna fits from 1.3 m, 1.9 m and 2.5 m, with a choice of rotation rates, 24 and 40 rpm.

Today, solid state radar transceivers (acting as transmitter and receiver sequentially) are replacing many magnetron based systems. For example, the Kelvin Hughes SharpEye™ radar transceiver does not contain a magnetron, which minimises routine maintenance while increasing reliability. Despite the apparent low peak power, especially in poor weather conditions, detection performance is superior to that of magnetron based radars. Clutter suppression is achieved in the SharpEye™ radar transceiver itself. Using patented pulse compression technology (discussed further in chapter 6) and radar return processing ensures the radar operator is presented with the targets and tracks on the display while minimising clutter from even severe rainstorms and high sea states, meeting modern day requirements of safety, navigation and collision avoidance at sea without heavy waveguide trunking, making the transceiver easy to fit on the mast. It also has a low peak transmission power of 200 W but is equivalent to a 30 kW magnetron, employing the first gallium nitride (GaN) power transistor technology.

Solid state radar, being compact and lightweight and with signal processing capabilities, are valuable for ocean, littoral operations, port and harbour security, and have found application in land-based mobile surveillance solutions in pulse Doppler radar modes such as the Kelvin Hughes SharpEye™ SxV, designed for vehicles on a single mast. The radome enclosed unit consists of an antenna providing 360-degree coverage.

From a Royal Naval perspective, SharpEye™ provides a suitable supersessor to Type 1006 and Type 1007 navigation radar currently in RN-service since 1988, the first solid state pulse Doppler navigation and situational awareness radar. Providing I and E/F band transceivers, a new generation of navigation and short range tactical surveillance radars exceeds the high performance requirements mandated by the IMO.

A comparison between a conventional I-band radar and a SharpEye™ transceiver system is given in figure 5.19, although a range of magnetron based naval radar is also being produced.

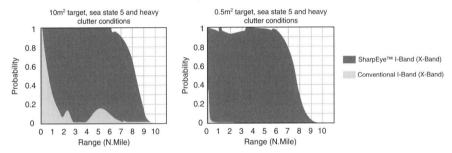

Figure 5.19: *Comparison between conventional I-band radar and a SharpEye™ system. (Image courtesy of Kelvin Hughes.)*

The I-band transceiver will replace existing Type 1007 and similar I-band navigation radars to deliver a significant improvement in sub-clutter visibility of about 30 dB, enabling targets with small RCS, typically 0.5 m^2, to be detected even in heavy seas and rain clutter – critical when dealing with modern 'asymmetric warfare threats' such as a small RIB filled with explosives.

5.16 Automatic Radar Plotting Aid (ARPA)

Last in our discussion is the *Automatic Radar Plotting Aid* (ARPA) – radar that can create tracks using radar contacts, calculating tracked objects' courses, speeds and Closest Points of Approach (CPA) and establishing if a danger of collision with another ship or landmass exists. ARPA development began following the sinking of the Italian liner SS *Andrea Doria*, which collided in dense fog off the east coast of the USA. The first commercially available ARPA, installed on the cargo liner MV *Taimyr* in 1969, was manufactured by Norcontrol, now a part of the naval manufacturer Kongsberg Maritime. ARPA-enabled radar is available for most small yachts as well. The International Maritime Organization (IMO) sets standards concerning the International Convention for the Safety Of Life At Sea (SOLAS) requirements for carrying suitable automated radar plotting aids. The primary function of ARPA is to improve the standard of collision avoidance at sea, to reduce the workload of observers by enabling them to automatically obtain information to perform as well with multiple targets as they can by manually plotting a single target.

A typical ARPA creates a presentation of the current situations and uses computer technology to predict future situations. ARPA assesses the risk of collision and enables operators to see proposed movements of their own ship. ARPA generally provides the following characteristics: true or relative motion radar presentation, automatic target and manual acquisition, digital readout of acquired targets to provide course, speed, range bearing and CPA, ability to display collision assessment information directly on the radar's Plan Position Indication (PPI) display, the ability to perform trial manoeuvres including course, speed and combined course speed changes, and automatic stabilisation for navigation.

ARPA processes radar information more rapidly than conventional radar. ARPA data is only as accurate as the data that comes from its data inputs. It should be noted that similar electronic tools exist for Air Traffic Control (ATC) in aviation as well. This and several other electronic aids, such as Voyage Data Recorders (VDR), Electronic Chart Display Information System (ECDIS) and Navtex, as well as ARPA, are discussed further in Reeds Marine Engineering and Technology Series, Volume 15: *Electronics, Navigational Aids and Radio Theory for Electrotechnical Officers* (ISBN 978–1–4081–7609–2).

Self-assessment questions

5.1   Explain how a pulsed radar measures target range.

5.2   A radar transmits a pulse of microwave energy that takes 2.5 ms to return an echo to the radar receiver. What is the range of the contact?

5.3   Which factors affect Maximum Detection Range? If the PRF of civilian navigation radar is 1500 pulses per second, what is the MUR? What is the *likely* MDR of the system going to be (2 significant figures)?

5.4   If a radar has a PRF of 2000 pulses per second, rotates ten times per minute and has a horizontal beam width of 1 degree, what is the *maximum* number of pulses (N) that will be expected to reflect from a small point target?

5.5   If a radar has a pulse resolution of 1 microsecond, what is the radar's range resolution (3 significant figures)? Explain how pulse duration is chosen for different types of radar.

5.6   State the Doppler effect and explain how it can be used to give target speed and target identification. If a radar transmits at a frequency of 10 GHz and a ship is moving directly towards the transmitting radar so that a frequency shift of 50 Hz is observed, what is the relative velocity between the two ships (1 decimal place)?

5.7   Explain what is meant by antenna gain. The isotropic intensity from a source is 5 W per square metre. The directional intensity from a source with a suitable reflector is 100 W per square metre. What is the gain of the antenna? What is the gain in decibels (2 decimal places)?

5.8   Draw a Cassegrain radar antenna, and label it. What would the typical radiation pattern from a Cassegrain radar antenna look like for both horizontal and vertical coverage?

5.9   Discuss the functions of a radome.

5.10   Draw the pulse delay ranging radar block diagram and explain the function of each component.

REFERENCES

[5.1] www.penleyradararchives.org.uk/history

[5.2] www.bbc.co.uk/dna/ww2/A591545

[5.3] *Stealth Warship Technology*, C Lavers (Reeds Marine Engineering and Technology Series, Volume 14, 2012, ISBN 978–1-4081–7552–1).

[5.4] www.weather.com/maps/maptype/dopplerradarusnational/index_large.html

[5.5] www.met.rdg.ac.uk/radar/camra.html

6

Radar Echo Signal Processing in a Real Operational Environment

'England expects that every man will do his duty'
The signal sent by Vice Admiral Horatio Nelson,
as the Battle of Trafalgar was about to
commence on 21 October 1805

6.1 Operating environment

We will now discuss the practical effects of operating a radar in a changing and complex multi-environment space (or battle space, from a military perspective) and consider signal processing aspects relevant to radar. These are general principles, which may be applied to other sensing systems.

For a radar operator, detection of real radar signals is of primary importance. We discuss detection of real radar signals against other echoes returned from the background (*clutter*) as well as random electromagnetic signals (*noise*), generated naturally in the atmosphere or receiver by man-made systems accidentally (or, in the military, by deliberate jamming), that confuse the radar picture. We introduce some simple common radar techniques that improve the radar picture so a radar operator can detect maritime vessels clearly despite these problems.

6.1.1. Processing functions

Radar training should include a full understanding of how radar works and the manual user display controls, which involve many basic functions as well as more sophisticated radar signal processing functions. If these are understood, the reader is likely to appreciate the advantages provided where most of these functions are performed in automatic or semi-automatic mode.

6.2 Threshold detection

Consider a typical voltage response of returning echoes and noise. A typical response is illustrated in figure 6.1.

Peaks marked A, B and D represent real contact echoes while C and E and other peaks represent noise. To show a contact, the receiver detector has a minimum threshold voltage for display and signals below this threshold are ignored. Correct threshold setting is critical to radar operation.

Figure 6.1 shows two settings of threshold voltage applied to the waveform. The first is a high setting and detects contacts B and D, but misses contact A, a *missed contact*. The second, lower threshold detects all three contacts, but gives a response to the noise peaks C and E, which are called *false alarms*.

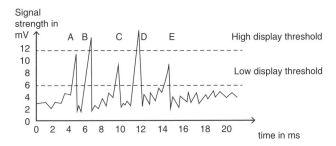

Figure 6.1: *Threshold levels for radar.*

If we set the radar detection threshold voltage *too* high, we will have a low or zero *False Alarm Rate* (FAR), but may miss radar contacts. On the other hand, if we set the threshold too low, we will not miss any radar contacts, but may have an unacceptably high FAR.

Example 6.1: In figure 6.1, how many false alarms are there with the low threshold? What voltage would correctly distinguish both targets from noise?
There are two false alarms. Approximately 10 V is the appropriate threshold.
Example 6.2: If the threshold value in figure 6.1 is 12 V, what will be the result?
The result will be two targets detected (B and D) but one missed contact (A), with no false alarms. Some radars maintain a Constant False Alarm Rate (CFAR) by measuring mean noise level and adjusting the threshold voltage.

6.3 Integration

Integration is a technique that helps improve detection in a noisy environment using time averaging. Signals and noise obtained from successive radar pulses are summed together. Real returns from several consecutive echoes are consistent and correlate with other real returns. Contact echoes remain in the same position and increase, but noise peaks are random in position, phase and amplitude and average out (figure 6.2). The longer the integration time (t_i), the higher the noise cancellation. Integration time is limited only by the time a radar looks at a contact (t_{ot}), so $t_i = t_{ot}$. The effect of integration on a radar operator's display is seen before and after respectively in figure 6.2.

Use of increased integration time presents little problem to a merchant ship operator. However, this will cause reduced update or 'refresh' rate of the display. If watching the target is paramount, as in the case of a fire control, a dedicated fire control radar is used.

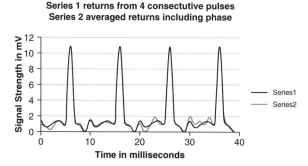

Figure 6.2: *Sequence of four consecutive pulse returns in blue, with averaging comparison in red.*

Averaging smooths out the overall response compared with single pulse returns, and consequently there appear to be less false alarm peaks present in averaged signals. Averaging in terms of amplitude, frequency and phase means the average level can be less than the actual simple summation of individual amplitudes, especially if noise phases are partially destructive (anti phase).

6.4 Clutter

A contact echo competes with unwanted echoes as well as noise. These echoes, or 'clutter', come from various sources: volume, precipitation and point clutter. Sea clutter is a form of surface clutter. Precipitation and chaff (a naval decoy) are both forms of volume clutter. Advanced radar can even detect point returns from individual birds or insects.

Clutter usually originates from small wavelets or droplets distributed over the sea surface or rain cloud. As clutter is a competing echo signal, increasing transmitter power cannot improve the signal to clutter ratio but does increase the signal to noise ratio.

6.4.1 Common maritime clutter sources

Sea clutter

Sea clutter increases as the wind increases and sea state worsens. Sea clutter is simply back reflection from a rough surface (diffuse reflection). It can be severe at short range and appear as an irregular shaped bright patch in the middle of the display. A smooth, calm surface tends to reflect waves in the forward direction, returning little radar energy, and is thus not a problem to the radar operator. Forward-looking satellite and high-altitude aircraft systems note increased microwave backscatter from a rough sea surface from satellite systems such as ERS (5.25 GHz), QuickScat (13.4 GHz) and ASCAT onboard MetOp since 2006 (5.25 GHz) [6.1].

Precipitation clutter

Precipitation clutter depends on: wave frequency, droplet diameter and precipitation rate. Rain gives larger echoes than snow or hail due to its different dielectric properties. Scattering sources are uniformly distributed throughout a precipitation volume, appearing as large, irregularly shaped echoes on the radar.

Clutter is a major problem in radar systems, for two reasons. It provides signals that compete with real contacts for the radar operator's attention and may mask weak real contacts (figure 6.3, where the signal to clutter ratio is less than 1.)

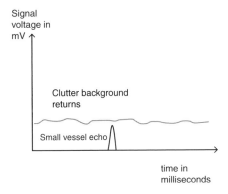

Figure 6.3: *Target not seen below clutter level.*

Clutter echoes may also be so big they can saturate a display. In the receiver they hide contacts, however strong, by providing maximum output at all times (figure 6.4). On a display, clutter and target echoes can reach the maximum display brightness and target echoes are not visible. Anti-clutter techniques overcome these problems by preventing clutter saturating the display.

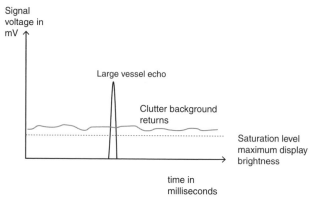

Figure 6.4: *Importance of avoiding saturation.*

6.5 Logarithmic amplification

Logarithmic amplification is a technique used to prevent receiver saturation. In linear amplification, output voltage (v_o) is the amplifier gain (G) multiplied by input voltage (v_i), or:

$$V_o = GV_i \hspace{3cm} \textbf{(eq 6.1)}$$

If input voltage is too large, a receiver saturates and output voltage will be constant, no matter how big the input voltage. Contacts that give much larger echoes compared to clutter may still be hidden (figure 6.4).

For example, if the echo strength of a ship is 50,000 times bigger than the echo strength of a small Rigid Inflatable Boat (RIB), you would not have 50,000 different brightness levels, and even if you did, you would never see the RIB on the display, compared with the ship.

Taking the logarithm of echo signal strength compresses the differences between the signal magnitudes, making it possible to display dissimilarly large and small echoes simultaneously.

This is achieved by extending receiver dynamic range with a suitable amplifier, where:

$$V_o = -G \log V_i \hspace{2cm} \textbf{(eq 6.2)} \text{ output in, say, mV.}$$

Example 6.2: If Vi = 10 mV and G = 20, calculate V_o for both a linear and logarithmic amplifier. If V_o = 150 mV is required for saturation to occur, which amplifier, if any, will be saturated?

$V_O = G\,V_i = 20 \times 10^{-3} = 0.2\,V$ saturated

$V_O = -G\,\log V_i = 20 \times \log(10 \times 10^{-3}) = 60$ mV

The logarithmic amplifier is unsaturated.

Logarithmic amplification prevents clutter from saturating a receiver. An anti-clutter switch can change between logarithmic and linear amplification. Linear amplification is still used as it gives a better response in low clutter environments with a good visual output.

6.6 Differentiation (anti-rain)

An anti-rain saturation technique is used against volume clutter (precipitation and weather fronts) that produces long clutter echoes on displays. Anti-rain uses a software algorithm that only responds to the echo's leading edge (figure 6.5).

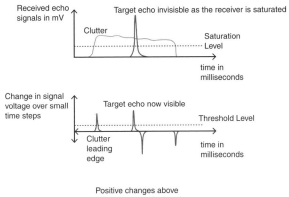

Figure 6.5: *Anti-rain clutter.*

Rectifying software following the differentiator removes the negative spike, so each echo becomes a positive spike on its leading edge. A contact is obscured by the clutter echo saturation, and isn't displayed. After differentiation and rectification (figure 6.6), the ship contact is visible.

Differentiation is an echo filter that eliminates long duration volume clutter and shows large contacts from within precipitation clutter (or, in the case of the military, chaff). Differentiation is often called *anti-clutter* (anti-rain) or Fast Time Constant (FTC) filtering.

Figure 6.6: *Volume clutter before (left) and after (right) differentiation, masked target is revealed.*

6.7 Swept gain

Sea clutter can easily saturate radar at close range. The effect diminishes rapidly with increased range due to the small Radar Cross Section (RCS) of the various clutter sources. We can overcome short range saturation by reducing receiver gain at close range and then increasing it to its maximum or normal value at the maximum range. Short range contacts, such as ships whose echoes are larger than the clutter signal, also become visible. Small boat contacts (e. g. RIBs) may be missed with reduced gain, but they were hidden by the clutter anyway. This is repeated for each pulse in turn and is illustrated in figure 6.7. This technique is called 'swept gain' or Sensitivity Time Control (STC). A radar operator's display is seen before and after swept gain respectively in figure 6.8.

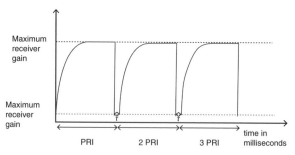

Figure 6.7: *Swept gain control.*

Figure 6.8: *Sea clutter before (left) and after (right) use of swept gain.*

6.8 Pulse compression

Good range resolution requires a short duration pulse. Long range detection requires a long duration pulse. It would seem to be impossible to achieve both good range resolution and long range detection simultaneously, as we would need both a short and long duration pulse simultaneously. However, there are more sophisticated radar pulse waveforms that can be used to overcome this conundrum [6.2].

We need to modify some property during a transmitted pulse that can be manipulated in the receiver. From the original wave definition in terms of amplitude and angle contribution, we see it is possible to vary amplitude or frequency (see Chapter 2). Varying amplitude is a poor choice as noise is an amplitude variation and long range echoes are likely to be very small and thus masked by noise. The best choice is to modify pulse frequency while keeping amplitude constant.

In pulse compression, we transmit a long pulse to obtain a long MDR and then compress it in the receiver to improve resolution, something not possible with conventional pulses. A radar transmits a Frequency Modulated (FM) pulse, so transmitted frequency rises during the pulse. If frequency increase is constant over the transmitted pulse, we obtain a linear FM pulse (figure 6.9, see plate section). Pulse amplitude remains constant. Cetaceans and bats make use of 'chirped' compression waveforms in the acoustic and ultrasonic parts of the sound spectrum respectively.

When an echo arrives back at a receiver, it is fed through a dispersive circuit where the output delay decreases with increasing frequency. Thus, higher frequency components at the end of the signal 'catch up' with lower frequency components at the start of the signal as they exit the receiver and are compressed on to its leading edge, somewhat similar to a tsunami wave in a littoral shallow water environment. In figure 6.9, we consider linear modulation as equivalent to a series of discrete frequency steps.

Range resolution improvement is seen for two overlapping FM echoes before and after the dispersive delay line (figure 6.10). Both signals are compressed on to their leading edges. As long as the leading edges are separated by a time greater than the compressed pulse length, both echoes are resolved separately.

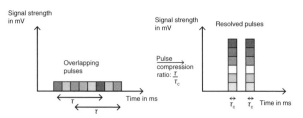

Figure 6.10: *Two adjacent ships resolved now with pulse compression.*

Operating in this mode increases peak power by a factor of: $\frac{T}{T_c}$, the *compression ratio*.

Example 6.4: If a transmitted pulse has a duration of 5 microseconds and a compressed pulse has a duration of 0.1 microseconds, what is the pulse compression ratio and compressed range resolution?

$$\text{Ratio} = \frac{T}{T_c} = \frac{5}{0.1} = 50$$

$$\text{Compressed range resolution} = \frac{c\,T}{2} = \frac{3 \times 10^8 \times 0.1 \times 10^{-6}}{2} = 15\ m$$

As the signal and the dispersive circuit are matched in the pulse compression radar, while real noise is random, the signal is compressed while noise is not. The compression ratio represents the increase in the receiver S/N ratio. A radar can get the range resolution advantage of a short duration pulse, but achieves the MDR advantage of a long pulse duration. Pulse compression is frequently used to improve long range surveillance radar range resolution performance.

6.9 Radiation hazards

Radiation from high power satellite and radars can present a significant radiation hazard. Radio and radar signals cause electric currents to flow in conductive objects they illuminate. Induced current size depends on the microwave frequency used, transmitter power, range, how good a conductor the object is, and how far waves penetrate the object. A flow of electric current can cause a heating effect. Waves at these frequencies can penetrate the body and heating can cause internal RF burns as well as skin (surface) damage. RF burns are usually caused by touching an antenna while it is transmitting and they can be painful and last for days. Burns and even electric shocks can result. Each maritime platform has its own procedure but general features of safe working on antennas include switching off all radio, radar and satcom equipment. It is advisable to remove the main DC supply fuses or keys, if accessible, to avoid accidental use of radio and other equipment while crew are aloft. All antenna switches should be moved to a position that disconnects the equipment from the antenna, and crew must put a sign on the radio or equipment stating that someone is working aloft. In all cases, one can limit exposure by reducing the time of exposure and staying as far from the RF source as possible. Prevent unauthorised access to all high voltage electrical equipment by locking all equipment cabinets where possible. Further safe antenna maintenance and antenna checks are detailed in *The Mariner's Guide to Marine Communications* by Ian Waugh (The Nautical Institute, ISBN 978–1-8700–7778–1).

6.10 Phased array

Phased array radars can be static or rotate. Static radar have no moving parts and use multiple antennas for 360 degree coverage mounted on ship superstructure they are lower than mast mounted radar, restricting search range due to weight. Fixed array accuracy falls off with angle. Rotating radar have moving parts and energy loss at rotating joints (passive only).

These variations lead to the classification of phased array radar into four categories (table 6.1).

	Static	Rotation
Active	Civilian Fraunhofer (FHR) Institute silicon-germanium (SiGe) maritime radar demonstrator, or German Sashen class frigate.	UK, Royal Navy, SAMPSON AESA radar.
Passive	US AN/SPY-1 naval radar.	Italian naval EMPAR (European Multifunction Phased Array Radar), a PESA rotating C-band multifunctional radar built by Selex ES.

Table 6.1: *Phased array radar types.*

6.10.1 Passive arrays

In passive arrays, one high power device produces energy fed to the antenna via waveguides. Sources are holes in the antenna. This process involves energy loss from waveguides and, if the antenna rotates, rotating flexible joint losses. If the power device fails, the radar fails catastrophically, i.e. all available power is lost.

6.10.2 Active arrays

Each hole in the antenna has a small solid state power supply behind it. Total radar antenna system power output is the integrated sum of all individual solid state emitter powers. Individual emitter failure results in *graceful* performance degradation with losses much less than passive antennas. A high percentage of active transmitter failure can still be tolerated due to the sophisticated digital software management system controlling transmission. The system is more complex and costly.

To acquire a target in space, a beam is steered horizontally and vertically, requiring a two-dimensional array with appropriate phase shifts respectively across the sub-array elements' arrangement (figure 6.11).

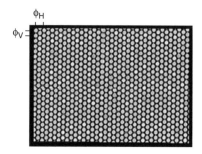

Figure 6.11: *Phase array with regular vertical and horizontal phase shifts between elements.*

Active Electronically Scanned Arrays (AESA) steer a radar beam electronically rather than by mechanical rotation. AESA use thousands of active microwave transmitting elements, through phase shifting, to steer a beam and are known as *phased array radars*. Such radars are used for Synthetic Aperture Radar (SAR) applications.

Since AESA radar can be steered rapidly in a fraction of a second, AESA radars allow ships or aircraft to use a single radar system for multiple applications, to all intents and purposes simultaneously, e.g. navigation, surface detection and tracking (finding ships), air detection and tracking (finding aircraft) and telemetry. For these reasons, AESA radars are multifunctional, able to perform many tasks at once.

AESAs are common on maritime naval platforms such as Ticonderoga (SPY-1) and TYPE 45 (SAMPSON). Both AESA and Passive Electronically Scanned Array (PESA) systems provide electronic steerable Multi-Functional Radar (MFR) performance. Civilian phased array technology lags behind military use, although it is available in meteorological applications [6.3]. PESA have been suggested for civil aviation applications, i.e. air traffic control.

6.10.3 Principle of operation

To explain how such radar works, consider two-source interference, with two sources separated by distance *much* greater than the wavelength, so the following interference pattern results (figure 6.12).

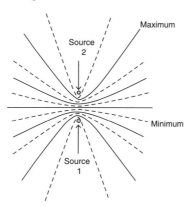

Figure 6.12: *Two-source interference pattern.*

If d, the source separation, is less than a wavelength, e.g. $d = 0.5\lambda$, with a suitable microwave reflector placed behind the two sources (figure 6.13) another interference radiated energy polar lobbing pattern results.

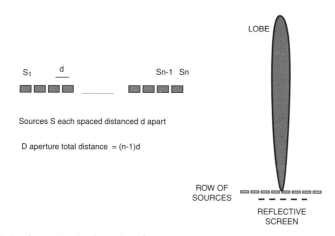

Figure 6.13: *Single constructive beam interference.*

Horizontal beam width is given by $a_H = 60\lambda/D$, where $d = D = \lambda/2$.

So $a_H = 60\lambda/(\lambda/2) = 120$ degrees, a beam too big for practical use. However, trying to produce a narrow beam from the formula with antenna length D increased, assuming λ held constant and source separation increased beyond wavelength λ, fails as complex multiple beams' interference pattern of maxima/minima returns (figure 6.12). Thus D must increase, but d remain less than λ. This apparent contradiction is achieved with a linear array of sources (figure 6.13).

As more sources are used, D increases but the d spacing is the same, giving smaller a_H values. Sources are in phase or with fixed phase difference between them. From figure 6.13 it is seen there is one less space than sources, so D = (n-1)d, where n is the linear array source number and

$$a_H = 60λ/[(n-1)d] \qquad\qquad \textbf{(eq 6.3)}$$

So a beam is formed from multiple source interference.

Example 6.5: A square phased array has 1600 sources spaced at regular intervals. If sources are spaced by 0.5λ, what is the horizontal radar beam width (2 decimal places)?

Total number of sources = n^2 where n is the number of sources in a row or column.

As n^2 = 1600

So n = 40

Using the equation: $a_H = 60λ/[(n-1)d]$ and

substituting:

$a_H = 60λ/[(40-1) 0.5λ] = 60/[(40-1) 0.5] = 120/[(40-1)] = 3.08$ degrees.

6.10.4 Linear array beam steering

If a time delay is introduced across an array so each source (moving right to left) radiates a little later than the previous one, the wavefront will tilt because energy travels further from the first source, and progressively less distance from subsequent successive sources (figure 6.14 left), giving rise to a squint angle for end-fed slotted navigation radar. If a time delay is then introduced from the opposite direction, the beam is rapidly tilted in the opposite direction (figure 6.14 right).

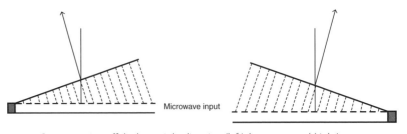

Figure 6.14: *Beam steering off the bore-sight direction (left), beam reversal (right).*

In real radar systems, phase differences between elements steer the beam as a time delay is equivalent to a phase change ϕ. The phase difference between elements is given by:

$\phi = 360d\ \sin\theta/\lambda$, where d is the source separation and θ the angle the beam is steered from its original position. Changing ϕ produces a beam in a particular angular direction.

    d is given in the form d = kλ with k between 0.5 and 0.9, so ϕ becomes:
    $\phi = 360\ k\ \sin\theta$ (**eq 6.4**).

As there is no mechanical antenna steering, beam steering is extremely fast. To acquire targets in space, beams are steered both horizontally and vertically with 2D arrays.

> **Example 6.6:** A radar operates with wavelength λ = 0.1m and is steered electronically, changing the phase angle of one emitter relative to the next adjacent emitter. If the beam is steered an angle of θ = 30° off the boresight, what phase angle ϕ must be applied between adjacent emitters?
> k = 0.5 (1 decimal place).
> Use the equation: $\phi = 360\ k\sin\theta$
> Substituting: $\phi = 360 \times 0.5 \times \sin90 = 90.0\ degrees$

6.10.5 Phased array radar advantages

Beam steering is very fast, typically several microseconds for different applications. Antennas produce several hundred beams per second and point to different locations in space. Some beams are used in search patterns. If a vessel is detected, other beams can track it. Further beams can track individual vessels or give guidance signals. MFR phased arrays perform multiple functions, which normally require use of several separate conventional radars.

Due to the large number of beams produced per second, data rate is high, yielding long MDR and high S/N ratio improvements. Antenna can be mounted rigidly on ship superstructure as there is no need for mechanical antenna movement. Recent phased arrays can also rotate on masts. The main phased array disadvantages are cost and complexity, which limit use on civilian vessels although they would be of great benefit. Phased array radars divide into two groups – passive and active – and may include many elements (e.g. Cobra Dane has 15,360 phase shifting elements per array) [6.4].

There are many naval maritime phased array radar systems, such as the RN SAMPSON radar (the Sea Viper anti air missile system on *Daring*-class destroyers). It

6.11 Synthetic aperture radar

Maritime radar was developed to detect ships at sea. There was a need to develop higher resolution systems and a technique enabling radar used on platforms at a great distance from a target, such as satellites. As a result, *Synthetic Aperture Radar* (SAR) was developed. SAR uses platform movement to make the antenna length appear longer than it is. Signal processing techniques produce the effects of a physically long antenna and hence narrow beam width without increasing actual antenna length (figure 6.15), hence the term *synthetic aperture*. This apparent increase in antenna size is achieved using the fact that *relative motion* between a radar set and a vessel at sea causes reflected signal frequencies from the ship to be slightly different to those transmitted. Using this frequency shift or *Doppler shift*, the effective antenna beam width can be reduced significantly so the radar along-track resolution is of comparable magnitude with the range resolution.

Using SAR techniques, it is possible to produce effective antenna beam widths of 0.1 degree or less. SAR historically was an aircraft radar platform mapping technique for generating high resolution maps of surface areas and terrain, but now aids in ship and oil rig platform identification at considerable distances.

So how do we get better angular resolution with a radar? The answer is to use a larger aperture *D* for the antenna. However, increasing antenna size to a length where we get improved resolution is impractical. Instead, we *simulate* an aperture length by moving a fixed length antenna along a path relative to the object.

Aircraft with SAR travelling at v metres per second will cover vT metres over the integration period T.

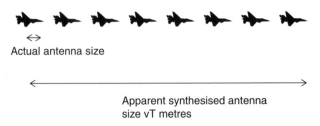

Figure 6.15: *SAR integration.*

SAR is viewed as a way to produce images at radar frequencies and takes advantage of the long range, low attenuation propagation properties of radar, and is less affected by atmospheric and weather conditions compared with other systems. SAR works in a similar fashion to phased array but instead of having a large number of parallel Transmitter/Receiver (T/R) emitter antennas, SAR can use a single T/R antenna, which transmits and receives independent signals over a common signal path. With phased array, beams are formed by switching emitters on at different times. In SAR, beam pattern is the result of the platform moving across the target with velocity v. SAR stores all radar returned signals as amplitudes and phases over a time period T. The digital SAR processor reconstructs a signal that would have been obtained from an antenna with aperture length $v \times T$, representing a larger aperture, and so higher resolution is achieved (figure 6.15).

Example 6.7: A maritime surveillance aircraft operates an AESA radar in SAR mode. The aircraft travels at velocity $v = 125$ ms^{-1} and transmits on a frequency of 18 GHz. The aircraft platform overflies the North Sea in a time period of $T = 50$ s. What is the synthesised aperture D length in metres (3 significant figures)?

SAR synthesised aperture length $= v \times T = 125 \times 50 = 6250$ m

---

SAR echoes from each transmitted pulse are recorded by the processor. As the platform moves over a ship, all the ship echoes for each transmitted pulse are recorded for the entire time the ship is in the beam.

Like a phase array, it is important for SAR signals to be *coherent*, i.e. they must be at a single frequency with constant phase difference. It is important, as the phase difference shouldn't fluctuate while the SAR scans across the target.

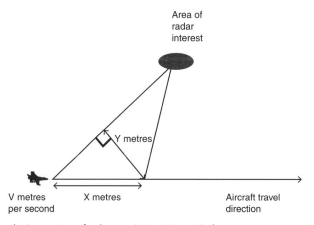

Figure 6.16: *Synthetic aperture of radar over integration period.*

For 3 cm waves, a 0.5 m antenna gives a 3.6 degree beam width from the equation: $a_H = \dfrac{60\lambda}{D} = \dfrac{60 \times 0.03}{0.5} = 3.6$ degrees.

If we observe an area of interest, with SAR acquiring a data track from a helicopter platform for one second at say V metres per second, we fly in the direct line some distance X, e.g. 400 metres (figure 6.15). Echoes received during this time are integrated, giving an apparent antenna much larger than the real thing. Allowing for the angle of observation, this could give Y = 200 m effective aperture. A 200 m antenna would in theory give a beam width of about 0.01 degrees. Hence, the attractiveness of this technique for satellites orbiting Earth at thousands of miles per hour.

The first civilian space based SAR (Seasat) was launched in 1978 by NASA and failed three months later but still produced data that proved very useful, demonstrating the value of satellite radar. Further details about Seasat can be found in a NASA publication [6.5]. Satellite SAR systems will form a major part of future maritime surveillance and ocean satellite missions. ESA's polar platform mission, ENVISAT, and the Earth Observing System (EOS), an international project, included SAR as a part of their payloads.

SAR is similar to Doppler processing. A given along-track co-ordinate on the sea surface has a unique time variation of Doppler frequency associated with it. As long as the return signal amplitude and phase are recorded, the Doppler component may be obtained. The process can be repeated for other values of along-track co-ordinates. With SAR imaging, a strip vertically below, at height H, if the real SAR antenna length is L, the beam width in radians is approximately $\beta \sim \lambda/L$, corresponding to a distance (arc length) $\lambda H/L$ along the track at the surface.

Looking at it the opposite way round, a given point on the ground is illuminated by the antenna only while the SAR radar travels this arc length distance; this is the maximum synthetic aperture length. The radar synthetic aperture angular resolution $\beta_{synth} = \lambda/(\text{max synthetic aperture length}) = \lambda/(\lambda H/L) \approx L/H$,

so the surface resolution $= \beta_{synth} \times H = (L/H) \times H$, giving a surface resolution of $\approx L$.

6.11.1 SAR imaging moving targets

If a target moves, the complex way in which data is processed results in a shift in apparent object position. As a radar approaches a stationary target, the Doppler shift decreases, reaching zero when the radar has achieved the same along-track position as the target. There is then no relative velocity component directly between the source and the target. If the target moves, a second Doppler shift is added to that due to platform motion. This means the Doppler shift sums to zero at a different along-track co-ordinate (different relative velocities) and the processor assigns this value to the along-track position.

Imaging radar has found wide applications in oceanography. Surface wave fields are imaged distinctly. Wave diffraction from coastal features and refraction by variations in bottom topography are often visible. Imaging radar is used to look at sea ice. Delineation of boundaries between ice floes and open water is relatively easy.

Across-track resolution is achieved with fairly conventional radar techniques such as pulse compression, and can be achieved with hardware capable of handling bandwidth of up to 500 MHz, leading to a resolution of about 0.5 m. Two SAR techniques used are *swath mapping*, where the ultimate resolution is limited to about half the real antenna length, as proven above, and spotlight. In *spotlight*, the radar beam is trained on a single patch of ground for several tens of seconds, and in principle centimetric resolution can be achieved. In either swath or spotlight imaging, higher resolution is gained.

6.11.2 Maritime monitoring with SAR

SAR can provide several different functions, with considerable advantages of being 24/7 (active rather than passive solar based electro-optical systems) and able to penetrate thick cloud and fog, providing high resolution sea surface imagery.

SAR modes include: vessel detection, oil spill detection, sea ice monitoring and iceberg detection.

Vessel detection

As shipping traffic grows, the pressure on the security of the world's oceans is increasing and near-real-time surveillance of shipping routes is essential to detect illegal and dangerous activity (e.g. illegal fishing, terrorism, narcotics traffic, immigration or piracy).

Oil spillage

Oil spillage information is required to restrict the environmental impact of accidents at sea, and to get distinct information about the source of oil pollution or leakages (to assist with audit trails and accountability in the case of prosecution with precise AIS GPS information). Radar satellite imagery is well suited to providing information on oil films on the sea surface. One such system now available is TerraSAR-X part of the Airbus Defence and Space Geo-Intelligence Programme.

Sea ice monitoring and iceberg detection

In high latitudes, strategic ice information provides the basis for routine ice monitoring, enabling planning of optimal ice free and economically viable routes through ice affected waters.

Surveillance radars are essential to a modern naval fleet and it is their ability to detect small and difficult targets in rough seas that is their most valuable maritime capability. Selex ES inherited the reputation of the lightweight Seaspray Mk1 in 1971 to provide the current Seaspray 5000 and 7000 series and the latest advanced system, Osprey. Developments of microwave phased array technology have taken us a long way since Seaspray's inception in the Cold War as a small ship's helicopter application on the first Lynx platform, then developed by Ferranti.

709 Naval Air Squadron was the first RN Lynx squadron, commissioned in 1978, leading to the introduction of the Seaspray radar type into over 40 frigates and destroyers. Observers now had a powerful sensor that gave them unprecedented situational awareness and the ability to detect small targets from sea and weather clutter. In search mode, the radar could see a wide forward-looking angle of 180 degrees, or narrower as required. Pulse durations were auto-selected by the radar according to the three possible range scales that were set. An identification mode also exploited transponder returns for target identification, with further navigational and search and rescue possibilities [6.6]. Seaspray led to development of the Blue Fox intercept radar for the RN Sea Harrier FRS Mk 7 carrier-borne fighters with pulse modulated capability. Blue Fox was replaced in the early 1990s by the then advanced Blue Vixen multimode pulse-Doppler radar.

In 1982, the same year Seaspray 'gained its spurs' in the South Atlantic conflict, the Edinburgh radar team sought to develop the next Seaspray generation, the Mk 3, which provided 360 degree all round coverage, with advanced processor techniques.

The Blue Kestrel radar, an I-band frequency agile radar, used an advanced Travelling Wave Tube (TWT) technology and pulse compression. Blue Kestrel proved its worth being able to pick up small radar contacts of Somali pirates across the vast expanse of the Indian ocean in the early 21st century.

In 1991, in the aftermath of the Cold War, GEC Ferranti privately funded a maritime surveillance radar portfolio, which led to the Seaspray 5000E Active Electronically Scanned Array (AESA) multimode surveillance radar (figure 6.17, see plate section).

The Seaspray 5000E radar is a compact, state-of-the-art AESA antenna to monitor air to surface coverage, and is installed in fixed wing and rotary wing platforms. It combines mechanical antenna scanning with electronic scanning to provide a wide range of capabilities from long range search to small target detection [6.7].

A real-time SAR processor was developed to provide target and vessel imaging capability (figure 6.18) and a range-Doppler map of the scene. Results were outstanding, with near-photographic quality resolution of surface water and land terrain and identifiable features to ranges approaching 200 nm.

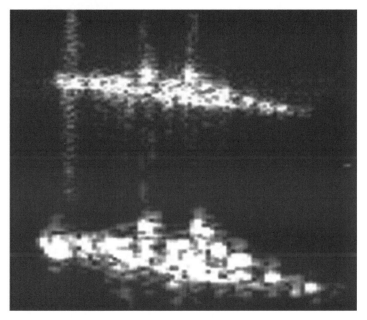

Figure 6.18: *ISAR image of two ships. Courtesy Leonardo Airborne and Space Systems Division.*

Doppler components generated are greater from the uppermost mast structures, but the majority of the superstructure is from lower bulkhead and hull components moving with smaller Doppler components.

The 1990s saw a maturing of solid state active phased array radars. Unlike mechanically scanned radar with a single transmitter and receiver, an AESA radar used gallium arsenide (GaAs) transmit–receiver modules and the array face provides independent control of phase and amplitude for multiple agile beams simultaneously.

The Seaspray 7000E multimode surveillance radar, with its ability to interleave modes and waveforms pulse to pulse, provided further improvements in target imaging with surface surveillance and weather detection simultaneously (figure 6.19, see plate section).

Combined with the Seaspray 7500E system with X-band operational installation (figure 6.20) and its SAR imagery, detection ranges of up to 320 nm are possible with a system weighing only 110 kg.

Figure 6.20: *The Seaspray 7000E scanner installed in aircraft. (Courtesy of Leonardo Airborne and Space Systems Division.)*

Seaspray provides a capability as yet only dreamed of in the civilian merchant and aviation arena.

Selex ES introduced the X-band 200 nm range Osprey multimode surveillance radar, providing wide azimuth and elevation electronically scanned (E-scan) fixed antenna with a compact, state-of-the-art processor and multi-channel receiver, providing land surveillance, small and low speed sea surface target identification, strip and spot SAR ground mapping, Moving Target Identification (MTI) and SART beacon detection, target imaging and classification (figure 6.21).

Figure 6.21: *Osprey multimode surveillance radar. (Courtesy of Leonardo Airborne and Space Systems Division.)*

Self-assessment questions

6.1   Explain why the detection threshold in a radar receiver must be carefully set.

6.2   Explain how integration can increase the S/N ratio of a radar echo signal.

6.3   With the aid of appropriate diagrams, describe volume and short range sea clutter and their likely appearance on a radar display.

6.4   With the aid of appropriate diagrams, explain how the effect of clutter signals on a radar receiver can be reduced using logarithmic amplification and differentiation.

6.5   If $V_i = 20$ mV and $G = 35$, calculate V_o for both a linear and logarithmic amplifier. If $V_o = 350$ mV is required for saturation to occur, which amplifier, if any, will be saturated and what will be the output of the linear and logarithmic amplifiers (2 significant figures)?

6.6   Explain with the aid of appropriate diagrams the advantages provided to a radar by using pulse compression.

6.7   A long range radar in ordinary mode transmits a pulse of 2.5 microseconds. In pulse compression mode, the radar transmits a pulse of 62.5 microseconds duration, but after passing through the receiver pulse compression filter has a duration of only 0.1 microseconds. What are the pulse compression ratio and the range resolution values for (i) ordinary mode and (ii) compressed range resolution (2 significant figures)?

6.8   State why radiation from radio and radar transmitters is a hazard to the human body, and list some of these hazards.

6.9   A long range search AESA operates with a wavelength $\lambda = 0.1$ metres and is steered electronically by changing the phase angle of one emitter relative to the next adjacent emitter. If the beam is steered an angle of $\theta = 30$ degrees off the boresight, what phase angle φ must be applied between adjacent emitters? $k = 0.5$ and phase angle $\Phi = 360\,k \sin \theta$ (1 decimal place).

6.10   A maritime surveillance aircraft operates an AESA radar working in SAR mode. The aircraft travels at a velocity $v = 210$ ms^{-1} and transmits on a frequency of 20 GHz. Using $c = f \times \lambda$, find the transmitted wavelength of the SAR (3 decimal places).

The aircraft platform overflies a certain area of the Mediterranean in a time period of $T = 55$ seconds. What is the length in metres of the synthesised aperture D (4 significant figures)?

REFERENCES

[6.1] 'Comparison of satellite microwave backscattering (ASCAT) and visible/near-infrared reflectances (PARASOL) for the estimation of aeolian aerodynamic roughness length in arid and semi-arid regions', C Prigent, C Jiménez and J Catherinot, Atmos. Meas. Tech., 5, (2012), pp. 2703–2712

[6.2] www.met.rdg.ac.uk/radar/ufam/pulsecomp.html

[6.3] www.nssl.noaa.gov/par/

[6.4] 'Phased Arrays and Radars – Past, Present and Future', E Brookner, *Microwave Journal* (31 January 2006).

[6.5] 'Seasat Views Oceans and Sea Ice With Synthetic-Aperture', Lee-Lueng F and B Holt, JPL Publications, NASA Jet Propulsion Laboratory (15 February 1982) earth.esa.int/documents/10174/1020083/Seasat_views_Oceans_Sea_Ice_with_SAR.pdf, pp. 81–120.

[6.6] *Seaspray Radar The First 40 years*, R Scott, Selex ES.

[6.7] *Looking Forward–60 years of Fire-control Radar*, R Scott, Selex ES.

7

Electro-Optical and Thermal Systems

'For those of you who are watching in black and white, the pink is next to the green.' Edwin Lowe, MBE (1920–2011)

7.1 Reasons for maritime imaging systems

Visibility is an important factor for safe navigation. Safe navigational visibility can be enhanced just by use of a bright searchlight (figure 7.1a, see plate section), or more detail may be provided by simple optical instruments such as a telescope (figure 7.1b, see plate section). Halogen metal halide lamps stand out well at sea with specific metal additions to the gas discharge, making them ideal for navigation in ice. Full intensity is possible after only 45 seconds, with average lifetime of lamps of up to 750 hours, ten times the typical lifetime of a halogen incandescent lamp, with ranges of up to 4000 m. Xenon lamps can provide ranges greater than 8000 m.

Poor weather conditions may require additional sensing abilities to provide safe surface detection. Thermal imaging cameras now allow captains to see in total darkness and view channel markers, shipping lane traffic, land and rock outcrops, other vessels and floating debris that could otherwise damage ships if undetected. Even small objects missed by radar can be clearly visible with a thermal imager.

Use of electro-optical (including thermal devices) has become routine in maritime applications for civilian and military applications. Electro-optics encompasses a broad range of spectral capabilities covering the visible spectrum, but now routinely (Near-infrared spectroscopy) NIR, (Mid-infrared spectroscopy) MIR and (Far-infrared spectroscopy) FIR. Systems have been developed for detection and imaging applications. Some utilise passive detection, others are active. Some cameras display in black and white 'palettes', while others use false colours. Some thermal (FIR) cameras now provide radiometric calibration (accurate remote temperature measurements). Prices of colour camera systems are now quite affordable, even for school applications.

Cameras can provide detailed imagery, especially valuable for maritime search and rescue (figure 7.2).

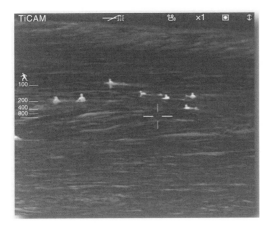

Figure 7.2: *TiCAM ® 600 camera imagery. (Courtesy of Thermoteknix Systems Ltd.)*

Thermal and Image Intensifiers are generically considered as 'night vision' but operate in different ways. The differences between them will be explored further in this chapter. Modern naval, merchant and pleasure boats routinely use low light level Close Circuit TeleVision (CCTV) for maritime and harbour monitoring. Night vision technology for night operations (visible and NIR imagery enhancement and thermal heat detection) are used regularly, besides civilian and military applications: lasers for ranging, at visible and invisible wavelengths. Thermal cameras are valuable for fire fighting, surveillance, iceberg detection, border security, predictive maintenance, electrical grid applications and many more applications.

There is an increasing number of suppliers of such technology worldwide and it is possible to obtain 'bespoke' multi-axis gyro-stabilised electro-optical systems composed of low light levels cameras with NIR, thermal and laser capabilities for applications such as Replenishment At Sea (RAS). As a consequence, it is important to examine understanding of the basic operational principles of key maritime sensors. Let us begin our discussion of night vision systems with one of the most frequently used electro-optical maritime sensors, the image intensifier.

7.2 Image intensifiers

Image Intensifiers (IIs) – low level light systems – can amplify light levels and NIR 30,000–100,000 times. They only amplify light and NIR, and don't work in complete darkness; however, even on dark nights there is nearly always a little scattered NIR starlight, allowing them to operate. The simplest types of II require no artificial light source and are 'passive' devices with low power consumption. Older three-stage

Cascade IIs are still available but suffer severely from bright lights in the FOV saturating displays. Various II generations exist, from Generation 0 Cascade passive II tubes and Generation 2 Micro Channel Plate (MCP) to Generation 3 devices providing high sensitivity and better response. The latest developments in night vision technology by comparison are focused around advances in the MCP II.

7.3 The Micro Channel Plate (MCP) Image Intensifier

The device is structured around a low pressure evacuated glass bulb so electrons emitted by the device during operation are 'driven' by a large potential difference (applied voltage between cathode and anode). One device is shown schematically in figure 7.3.

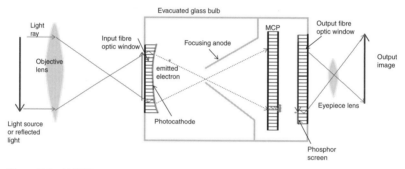

Figure 7.3a: *MCP II in cross section.*

Figure 7.3b: *MCP II.*

An objective lens transmits an image to the fibre optic window. Light is guided through individual optical fibres without spreading out until it reaches the surface of the *photocathode*, a device that converts light (photons) into an output of electrons (at the cathode). It is an evacuated glass structure with low pressure so emitted electrons are accelerated around the device towards the conical positive anode. The conical anode containing the MCP and the phosphor output screen is seen in figure 7.5a. The complete glass-encapsulated II is seen in figure 7.5b.

Figure 7.5a: *Anode. 7.5b: Encapsulated MCP tube.*

At the photocathode, the *photoelectric effect* emits electrons, a process first explained by Albert Einstein (1879–1955) in 1904. Electrons are only emitted from the cathode if they exceed a minimum energy or work (electron binding energy), quoted in electron Volts (eV) and given the symbol phi (ϕ). The number of emitted electrons depends on the overall radiation intensity striking the photocathode.

If the anode is held at a potential of +15,000V, emitted electrons accelerate in straight lines towards the anode but may pass instead through the entry cone of the wrapped anode (figure 7.5a), entering the MCP where electrons are multiplied by secondary emission. When many energetic electrons leave the MCP they hit a phosphor screen, light is emitted, and an image appears on the screen, through the inverse photoelectric effect.

7.4 Photoelectric emission of electrons

Light or shorter wavelength radiation can cause electron emission. Emission is explained if light consists of energy packets or *photons*, rather than being continuous waves. Photon energy (E) is given by $E = hf$, where 'h' is Planck's constant (6.6×10^{-34} Js) and 'f' is the wave frequency. High frequency (short wavelength) photons carry the most energy. If photon energy is sufficiently larger than the photocathode's *work function* (ϕ) (a material parameter), it ejects electrons and gives them some Kinetic Energy (KE), i.e.:

$$E = hf = \phi + (½ \, mv^2)_{max} \qquad \textbf{(eq 7.1)}$$

where 'm' is the mass of the electron and 'v' its speed.

Example 7.1: If an incident photon frequency is 650 nm, and the work function of the electron emitting surface is 1.5 eV, what is the maximum kinetic energy of the emitted electrons in eV (2 decimal places)?

$$E = hf = \frac{6.6 \; 10^{-34} \times 3 \times 10^8}{650 \; 10^{-9}} = \frac{3.0554 \; 10^{-27}}{1.6 \; 10^{-19}} = 1.91 \; eV$$

So KE = 1.91 − 1.5 = 0.41eV

As photon frequency is decreased, photon energy and the KE of the emitted electron will fall. Finally, a frequency is reached where an electron is just emitted with 0 KE. This frequency, f_o, is the threshold frequency and given by:

$$E_o = hf_o = \phi \qquad \textbf{(eq 7.2)}$$

Below this frequency, the incident radiation simply doesn't have enough energy to liberate an electron with this process. Photoelectric emission is used in image intensifiers and photomultipliers. Metal oxides such as tungsten and barium oxides have very low work functions, but nonetheless intensifiers only operate to about 1.1 microns wavelengths, and only exploit a small part of the NIR spectrum.

7.5 Secondary emission of electrons

Certain materials when bombarded by 'primary' electrons emit *secondary electrons*. The material's secondary emission coefficient δ is defined by:

$$\delta = \frac{Number \; of \; secondary \; electrons}{Number \; of \; incident \; electrons} \qquad \textbf{(eq 7.3)}$$

where δ depends on the specific material and electron Kinetic Energy (KE).

Example 7.2: If 15 electrons are emitted on average for every three incident electrons, what is the secondary electron emission coefficient?

Using $\delta = \dfrac{Number \; of \; secondary \; electrons}{Number \; of \; incident \; electrons}$ $\delta = \dfrac{15}{3} = 5$

An MCP consists of many tiny channels running parallel to each other between the plate surfaces. Each channel acts as an electron multiplier with a diameter of about 10 microns (10×10^{-6}m) and releases several electrons for every electron striking it, i.e. $\delta > 1$.

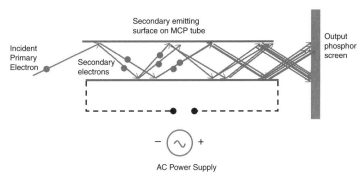

Figure 7.6: *One of millions of MCP tubes.*

In each half cycle of the AC power supply, the split ring electrode coating down the channels means a primary electron striking the anode in its first half cycle emits two secondary electrons. These are now attracted and accelerated towards the opposite half of the split ring, which becomes the new anode.

The MCP II's photocathode is sensitive to visible and NIR, so the II can use the NIR present in the night sky. The final image colour seen in the instrument depends upon the phosphor type. A green phosphor is used as the human eye is most sensitive to the colour green. Image intensifiers have been trialled for daytime use as NIR penetrates dusk and haze better than visible light, providing greater horizon detection range and imaging under these conditions. Bright light sources in the Field of vision (FOV) are suppressed without affecting the whole image. Channels that accept large numbers of 'primary' electrons saturate without affecting other adjacent channels. MCP IIs also have advantages for civilian and military use, as they are small and light, ideal for lightweight aviation sights.

7.6 Maritime II applications

There are many maritime applications that benefit from intensifiers, such as: harbour and port surveillance, driving sights, flying goggles etc. They can be fitted to TV cameras or low light level CCTV (figure 7.7, see plate section).

Because IIs use visible and NIR radiation, they cannot 'see' through fog, mist or smoke. From a military operational perspective, it is easier to operate devices under a full moon and with clear skies with good weather conditions. However, from a merchant shipping and civilian point of view, 'ideal' conditions may not be possible.

Typical range values for a low light level TV system used against a small boat might be:

Twilight/full moon 15km
Starlight 4km

It is worth commenting that detection, recognition and identification are not the same thing. It may be just possible to detect a vessel at a 20 km range but there may not be enough light to recognise the vessel type, let alone identify the individual vessel. Generally, one needs improved light levels or to approach the vessel to obtain a better picture.

7.7 Colour Image Intensifier (CII)

Several generations of intensifier have been developed, improving the overall imaging capabilities of these devices. Recent developments look at improving image recognition with false colour representation, in so-called Colour Image Intensifiers (CII). Helmet-mounted systems for aviators' night flying goggles are being developed incorporating two II tubes for stereo viewing, particularly useful in collision avoidance. Many displays rely on a conventional transparent helmet with a central portion portraying a night enhanced image and a smaller section for data and text display (figure 7.8). Systems work with standard AA size 1.5 V batteries, and contain a video-display power and signal controls belt pack.

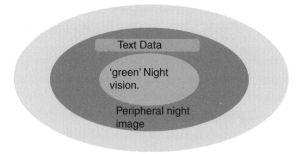

Figure 7.8: *Night vision display centred, peripheral night vision and text display position indicated.*

False colour systems take spectral differences and convert them into different visual colours. Colour contains spectral information while Black and White (B/W) doesn't, having only grey shade to differentiate levels. Figure 7.9 (see plate section) illustrates an identical scene with two objects in the FOV with almost the same overall intensity as the background, but much more NIR than the background.

A B/W camera has only one sensor sensitive to intensity differences. A B/W monitor has one phosphor, rendering intensity differences as greyscale. A colour monitor has three phosphors – red, green and blue – so objects with a different spectra provide higher contrast because CIIs have two sensors, i.e.:

(i)   A red sensitive sensor where the II tube outputs green, and

(ii) An NIR sensitive sensor where the second II tube outputs visible red.

Both images are viewed on the same monitor screen overlapped. CIIs have many potential applications, from military to humanitarian roles: mine detection, observation, driving, and for Unmanned Aerial Vehicles (UAVs). Key features of such systems exhibit faster recognition, fewer recognition errors, better depth perception, real-time contrast enhancement, and the ability to recognise different terrain such as grass or sand.

For example, consider figure 7.10 (see plate section): the left-hand image is created by a single phosphor in greyscale for all detected wavelengths. Distinguishing one vegetation type from another isn't easy as different wavelengths have the same overall intensity. The middle picture with full colour (conventional camera) gives imagery with red, green and blue, optimising spectral resolution. The right-hand image is output of a dual colour display – one sensor detects red and outputs red, the other detects green and outputs green. Blue is not detected and appears black. With a modern II system, NIR is displayed red while red (or green) is detected and displayed green, so object recognition is improved.

Further developments to night vision technology will likely include full digital data acquisition, in-device signal processing and wireless data linking, so a viewed scene can be sent to other rescue workers in a disaster area. Night vision goggles require wider FOV, are lighter, and should see beyond 1100 nm and have compatibility with other systems and the ability to import/export imagery.

II systems are light, small and simple to use, repair and replace. They provide the ability to detect and identify targets at short to long range, are complementary with Thermal IR and can work underwater at close range. I^2 systems are rugged and robust and have lower power consumption than thermal systems with 'normal' batteries.

7.8 Future wide-angle night vision systems

Most commercial night vision systems provide limited 40 degree FOV, and less than the stereoscopic human vision limit (nearly 180 degrees in the horizontal for overlapping eyes). The view of a current II system is typical of figure 7.11 (left – see plate section). Wider FOV systems benefit civilian and military applications. Current generation II systems have been developed with two or three overlapping II tubes (figure 7.11, right). Future systems will have greater than 100 degree FOV, providing superior situational awareness. Without scanning the head side to side, the aircraft in figure 7.11 (right) will be missed with the II used in figure 7.11 (left). Obstructions such as pylons, etc., become readily apparent with a wider FOV.

Current generation 3 gallium arsenide (GaAs) II tube technology is enhanced by gating, which means an II tube may be switched on and off in a controlled way. An electronically gated II tube functions like a camera shutter, allowing images to pass through when the electronic 'gate' is enabled. Gating durations can be very short (nanoseconds or even picoseconds). This makes gated II tubes ideal for use where very short duration events must be photographed or saturation conditions could be a problem, e.g. with nearby harbour lighting or in a building with some lighting in a corridor. Further II developments are followed elsewhere [7.1].

7.9 Lasers for marine applications

The advent of the LASER (Light Amplification by the Stimulated Emission of Radiation) has revolutionised digital telecommunications, sensing, medical treatment and precision engineering, operated across the electromagnetic spectrum. Lasers are well suited to these tasks.

7.9.1 Laser theory

Lasers have good spatial coherence in that output, being a diffraction limited narrow beam, is focused into a very small spot, achieving high irradiance and powers, even at considerable range. A laser consists of a *gain medium*, a mechanism to supply energy to it and something to provide optical feedback. The gain medium is a material with properties that allow it to amplify light by stimulated emission. Light of a specific wavelength passing through the gain medium is amplified. For the gain medium to amplify light, it must be supplied with energy, through the process of pumping. Energy is typically supplied as an electrical current, or light. Pump light can be provided by a flash lamp or another laser.

In 1917, Einstein predicted that excited atoms may emit radiation at a frequency equal to that of an emitted photon. This process of *stimulated emission* is the cornerstone of laser operation (figure 7.12). In stimulated emission, a photon is emitted with an energy difference hf_{if} between an initial excited energy state ε_i and another final lower energy state ε_f, ie:

$\varepsilon_i - \varepsilon_f = hf_{if}$, where h is Planck's constant   **(eq 7.4)**

Light emission (and absorption) is quantised. Energy is only absorbed in discrete sized steps or 'quanta' because energy content of an atom or molecule is limited to only a discrete numbers of levels, so generally:

$E_n - E_{n-1} = hf$ **(eq 7.5)**

where E_n and E_{n-1} are next nearest related energy levels.

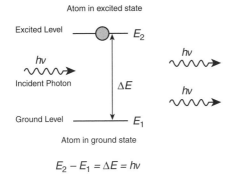

Atom in excited state

Excited Level ── E_2

hv
Incident Photon ΔE

hv

hv

hv

Ground Level ── E_1

Atom in ground state

$E_2 - E_1 = \Delta E = hv$

Figure 7.12: *Lasing difference showing the quantisation of energy between two energy levels.*

Example 7.3: The initial energy state of excited atoms in the laser medium is 2.5 eV and the final energy state of the laser medium is 2.0 eV. What is the frequency of emitted laser radiation (3 significant figures)?

$E_n - E_{n-1} = hf$

$2.5 - 2.0 \text{ eV} = hf$

$0.5 \text{ } eV = 0.5 \times 1.6 \times 10^{-19} J = 8 \times 10^{-20} J$

$E = hf = 8 \times 10^{-20} J$

$f = \dfrac{8 \times 10^{-20}}{6.62 \times 10^{-34}} = 1.21 \times 10^{14} Hz$

Emitted (stimulated) photons have the same polarisation and phase, and propagate in the same direction as the stimulating photon due to the highly polished and parallel cavity ends. One face is highly reflective, while the other face is partially reflective. Stimulation can be achieved, as it was in the first ruby laser, with a 'flash lamp' to stimulate a *population inversion* in the laser medium, with more atoms in an excited state than in the ground state. An increasing wave of stimulated photons sweeps back and forwards along the cavity until it reaches sufficient intensity to escape from the partially reflective end of the laser cavity. Photons emitted not perpendicular to the surfaces of the polished faces are lost out of the cavity sides, which explains the highly collimated (parallel) nature of the exit beam (figure 7.13). The most common lasers use feedback from an optical cavity with a pair of mirrors either end of the gain medium.

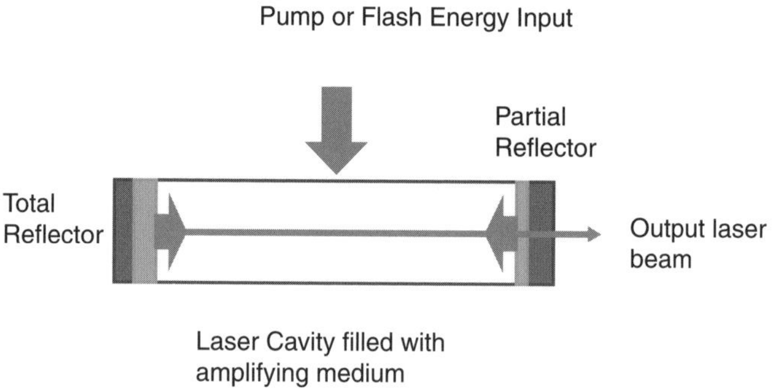

Figure 7.13: *Laser cavity.*

Emitted laser wavelength depends on the characteristic energy level transitions and occurs over very narrow wavelength ranges, approximately 0.1 nm. It is common to modify the spectral wavelength by adding one or more 'rare earth' ions. If focused, short laser pulses may yield intensities up to 10^{21}Wm^{-2}, compared with an oxyacetylene flame, which is only 10^7Wm^{-2}. A CO_2 laser can now burn through several mm of stainless steel plate in less than 10 seconds.

7.10 Laser classes

Laser products are labelled with a safety class number, which identifies how dangerous a laser is.

Different countries have different laser classification standards. The British Standard sets out seven classes of laser: 1, 1M, 2, 2M, 3R, 3B and 4. The higher the class, the greater the laser radiation hazard. Class 4 lasers are high power devices, usually needing a mains power supply and used for specific applications in research, medicine and industry. They are also used in the entertainment industry. Class 4 lasers are not designed to be used as laser pointers. It is strongly advised that the full reference is referred to for a clear understanding of your risks. The following laser classification scheme is taken from BS EN 60825–1: 2007. Safety of laser products: Part 1. Equipment classification and requirements. British Standards Institution, London [7.2].

Class 1 output power is below the level at which it is believed eye damage will occur. Beam exposure of a Class 1 laser will not result in eye injury. It is regarded as safe, because the light is contained in an enclosure, e.g. CD players or those used in supermarkets.

Class 1M lasers produce either a highly divergent beam or a large diameter beam. These laser products can be harmful to the eye if the beam is viewed using magnifying optical instruments. Some optical fibre communication systems are Class 1M laser products.

Class 2 lasers are limited to a maximum output power of 1mW and the beam must have wavelength 400–700 nm and is safe during normal use, as the human eye blink reflex will prevent damage. However, deliberate exposure to the laser beam may not be safe. Some laser pointers and barcode scanners are Class 2 laser products.

Class 2M lasers produce a highly divergent beam or a large diameter beam in the visible wavelength range 400–700 nm. Only a small part of the whole laser beam can enter the eye and they are limited to 1mW, similar to a Class 2 laser product. However, these products can be harmful to the eye if the beam is viewed using magnifying optical instruments for long periods of time.

Class 3R lasers are higher powered devices than Class 1 and 2 and have maximum output power of 5 mW. Such laser beams exceed the maximum permissible exposure for accidental viewing and can potentially cause eye injuries, although the risk of injury is low. Staring into such a beam for several seconds is quite likely to damage the retina.

Class 3B lasers may output power up to 500 mW and have enough power to cause eye injury from the direct beam or reflections. The higher the device output power the greater the risk of injury. Class 3B lasers are considered hazardous to the eye.

However, the extent and severity of eye injury from exposure to laser radiation from Class 3B lasers depends on several factors including the radiant power entering the eye and the exposure duration. Class 3B lasers are not suitable for general use as they can cause immediate eye damage on exposure.

Class 4 lasers are the most dangerous. They have output power greater than 500 mW and can burn skin; even indirectly scattered light has been known to cause eye and skin damage (https://www.gov.uk/government/publications/laser-radiation-safety-advice/laser-radiation-safety-advice). There is no upper restriction on output power. Class 4 lasers are a fire hazard.

A laser is a concentrated beam of radiation, which may not always be visible and can be dangerous, regardless of whether it is viewed directly or is a reflection from a smooth surface. Even with a low power laser system, never view the beam directly. Maintenance workers who examine inside systems are usually the most at risk. They should be trained and follow a work system that includes the use of protective eyewear; high power lasers are normally inside a safety interlocked enclosure. Identify and clearly mark all hazard areas. Maintain equipment regularly to minimise exposure to laser radiation. Instruct crew about the dangers of laser radiation and the precautions that should be taken.

7.11 Types of laser

Lasers fall into a distinct man-made classification of categories based on the lasing medium (material) used. The most common types are solid state, gas, semiconductor diode, liquid and optical: fibre, planar and channel.

7.11.1 Solid state laser

The Ruby laser, demonstrated in 1960 by Theodore Maiman (1927–2007), was the first of the solid state class of lasers, whose general output wavelengths now range typically between 400 nm and 2500 nm and beyond this wavelength as well, opening the way to new generations of electro-optical devices for sensing and telecommunications applications [7.3 and 7.4].

There have been many solid state laser developments including the 'laser comb', permitting precise emission at multiple wavelengths, whose Nobel prize-winning inventor was Thomas Hänsch [7.5].

7.11.2 Gas laser

The laser cavity is filled with gas, the active medium, which operates from the FIR to the UV. Typical gas lasers include: helium-neon (red 632.8 nm), argon, krypton, carbon dioxide (CO_2) and molecular nitrogen. The CO_2 laser, used for range finding, operates at 10.6 microns in the FIR with high kW output power levels possible.

7.11.3 Semiconductor diode laser

The first semiconductor laser (gallium arsenide) dates from 1962. Semiconductor lasers, made from layered materials, span from the UV to the FIR. Advantages of these lasers are their small size, precise fabrication and mass production. They are used for a wide range of applications: laser heads in CD players, checkout tills and light pointers.

7.11.4 Liquid laser

The first liquid laser used organo-metallic compounds such as rhodium and titanium complexes. Less toxic non-organic liquids have been discovered and fluorescent dye solutions such as rhodamine are commonly used. These liquid lasers can be tuned over a wide range of wavelengths (typically 100 nm tuning).

7.11.5 Optical fibre laser

Significant advances have been made since first demonstration of the single mode optical fibre laser in 1985 [7.6]. Since then, development has been rapid, with many doping materials exhibiting laser properties in fibres. The principal advantage of optical fibres is that they are convenient for conveying large amounts of information over long distances for telecommunications (eg transatlantic optical relays) and are thus an integral part of modern communications systems. There are three main optical fibre communications 'windows': 850 nm, for Local Area Networks (LANs), 1300 nm and 1550 nm for long distance communications. The optical fibre laser enabled engineers to amplify light pulses without external amplifiers, minimising introduction of further electronic noise.

7.12 Military and civilian maritime uses of lasers

Lasers are becoming common in civilian as well as military systems, as lasers become more robust, cheaper and smaller. Lasers now offer an alternative to radar for applications such as: ranging and tracking, underwater mapping, combined surveillance systems and target designators.

7.12.1 Laser range finding and maintaining ships' proximity in RAS

A laser range finder uses a laser beam to determine distance to an object in the same way as a microwave radar system does for navigation radar. At optical frequencies, the wavelength is short and precise distance measurements can be made. Some systems with very short and well-shaped laser pulses can range an object to a few millimetres. Range finders provide an exact distance for civilian engineering applications. Laser range finders are generally considered to be eye safe, meeting Class 1 safety requirements. Several laser systems maintain an exact distance between vessels, necessary for RAS operations. One system, Laser Atlanta SPROX (Ship PROXimity) @™, is a recognised 'laser distance measuring kit', ensuring ships' crews maintain a safe distance during underway replenishment and RAS, approved by the US Chief of Naval Operations and NATO. SPROX measures and displays ship distances in a fraction of a second on an easy-to-read waterproof. Portable equipment easily withstands rugged outdoor use and lowers the risk of dangerous manoeuvres while at sea, ideal for operations in all weather conditions, and can measure ranges up to 5000 metres to an accuracy of ± 0.5 metre.

Another laser-based reference system currently available is SpotTrack, which withstands harsh weather conditions, minimising maintenance available from Kongsberg. SpotTrack provides a reference system for relative positioning in Dynamic Positioning (DP) operations and is extremely accurate at short range and ideal for 3D positioning and vessel traffic monitoring.

7.12.2 Laser ranging and tracking

A laser is a near ideal system for tracking targets at very low altitudes as they have a narrow pencil beam and narrow FOV. The CO_2 laser is especially suitable as its wavelength lies in the 8–12 microns FIR window used by most thermal imaging systems and can even use the same optics as the thermal imager.

7.12.3 Lasers for underwater detection

Blue-green lasers can penetrate sea water to 2 km for underwater mapping applications and have been trialled for underwater mine detection [7.7]. Underwater lasers are attractive as they can be deployed from fixed wing and rotary wing aircraft on the end of a tether, ideal for surveying or directly down to the water surface. Complex electronic gating is used to prevent bottom echoes confusing the picture and the range achieved depends on water turbidity.

7.12.4 Combined surveillance systems

Laser range finders can be used in combination with other electro-optic devices (thermal imagers, low light TV Cameras or image intensifiers) to detect and track low level targets or hazards. Detection and tracking can be performed passively by electro-optic devices while the laser supplies target range. Having such narrow beam width, they are extremely difficult to detect and the entire system may be considered passive.

7.12.5 Satellite laser communications

The future of satellite communications will incorporate laser communications. In 2014, the concept was proven by NASA with its Laser Communications Relay Demonstration (LCRD), in which a lunar orbiter successfully sent and received data via a laser link to a ground station on Earth. Laser based systems are attractive as they offer a high rate of data transfer, provide digital transmission with higher accuracy and less data loss, and being narrow beam width are harder to intercept – a valuable military benefit. Laser satellite systems will be important for wide bandwidth, long range, spaced based applications.

7.13 Infrared emission and transmission

Infrared techniques have arisen to provide devices for imaging in the absence of visible and NIR, and overcome problems encountered with scattering in typical maritime weather such as mist and fog, which impact on collision avoidance in restricted visibility.

This resulted in development of sensitive devices for passive target surveillance and observation. Since small temperature changes create large IR emission changes, it is possible to detect objects that differ from each other slightly in temperature. There are bands in the IR spectrum where transmitted radiation is only weakly attenuated by the atmosphere, so devices operating in these regions achieve long range detection. Sensing of thermal or heat radiation has been used by mankind for thousands of years – there is anecdotal evidence that physicians to Egyptian pharaohs used their hands to 'feel' for temperature rise across the body of a sick pharaoh. Simple infrared temperature indicators (pyrometers) are readily available from suppliers such as RS, adjusted to provide the correct 'emissivity' value to be added, discussed later.

However, modern interest in infrared or heat fell to the work of Frederick Herschel (1738–1822), a German-born British composer and astronomer. On 11 February

1800, Herschel was testing filters to observe sun spots. When using a red filter, he found there was a lot of heat produced. Wanting to quantify the heating effect, Herschel discovered radiation in sunlight by passing it through a prism and holding a thermometer just beyond the red end of the visible spectrum. This thermometer was intended to be a control to measure ambient air temperature in the room. Imagine his surprise when it showed a higher temperature than the visible spectrum. [7.8]

7.14 The IR spectrum

Infra-Red (IR) or thermal wavelengths are measured in 'microns' where 1 micron or 1 μm = 10^{-6} m. Because of the large IR spectrum, it is divided for sensing into three regions: the NIR (0.70–3.0 μm), MIR (3.0–6.0 μm) and FIR (6–15 μm). The NIR band is regarded as a 'reflective' band as most things in this band reflect light predominantly. FIR and MIR are considered 'emissive' bands as quite ordinary room temperature objects radiate significant heat. Protective clothing may be needed to reduce warming, burning and irritation of skin from some hot surface. Eye protection with suitable filters should be worn to avoid discomfort and is essential with some infrared sources, such as lasers. Possible infrared effects include reddening of the skin, burns and cataracts.

7.15 Infrared emission laws

Infrared radiation (or heat) is emitted by all objects at temperatures above so-called absolute zero (−273 °C). The most common temperature scale is probably the centigrade scale (°C) developed by Anders Celsius (1701–1744), a Swedish astronomer, physicist and mathematician, who was professor of astronomy at Uppsala University (1730–1744). He made a rather arbitrary choice, focusing as he did on the range between the melting point of water (0 °C) and the boiling point of water (100 °C). The difference in height of a column of mercury between these two states, divided into 100 equal steps, provides a quantified and reproducible measure of temperature between the lower and upper limits.

However, energy is also related to temperature and one could get the erroneous idea that an object at 0 °C has *no* energy as energy E is related to temperature. In fact, a block of melting ice has a considerable amount of energy, if only it could be related to a true absolute energy scale. There is in fact such an energy scale, the Kelvin scale, developed by William Thomson, 1st Baron Kelvin (1824–1907). He had extensive maritime interests and was most noted for his work on the mariner's compass, which had previously been limited in reliability. Absolute temperature is given in units of kelvin (K). While the existence of a lower limit to temperature (absolute zero) was known prior to his work, Lord Kelvin is known for determining its correct value as very close to $-273.15°C$ or $-459.67°F$. In 1892, in recognition of his achievements in thermodynamics, he became Baron Kelvin of Largs. He was the first British scientist elevated to the House of Lords.

Returning to our block of melting ice, if all the energy from the ice, except its fluctuating quantum mechanical energy, could be removed, it would eventually fall to a very low temperature indeed, in fact close to $-273.15\ °C$ (2 decimal places). On the Kelvin temperature scale, a 1 °C drop is equivalent to a 1 K fall in temperature and $-273\ °C$ at the bottom of the temperature scale is given the value zero or 0 kelvin.

To convert between degrees centigrade (°C) and kelvin (K) is simple – we just add 273.15 to convert to kelvin, and subtract 273.15 to convert to °C.

Generally:
Kelvin temperature = centigrade temperature + 273.15   (**eq 7.6**)

Example 7.4: If an object is at 23 degrees centigrade, what is the object's absolute temperature in kelvin (2 decimal places)?
Kelvin temperature = 23 + 273.15 = 296.15 K

However, the amount emitted, and its wavelength distribution, depends on the absolute temperature (T) and on the 'emissivity' (ε) of a body. Emissivity tells us how good an emitter an object is. Emissivity is the ratio of actual emission from a unit surface area of a body to that from a perfect radiating source or 'black body' at the same temperature, where a black body has an emissivity of 1. Many real objects resemble a black body in their emission. Generally, the emissivity of a body is considered to be a constant of the emitting surface, regardless of temperature. However, this is not always the case.

Example 7.5: If a black body emits 100 W and a test material emits 92 W, what is the material emissivity (2 decimal places)?

$$\text{Emissivity of the material} = \frac{\text{Test material emission}}{\text{Black body emission}} = \frac{92}{100} = 0.92$$

Some typical examples of emissivity are given in table 7.1.

Material	Emissivity value
Smooth ice	0.97
Steel alloy type 301	0.27
Wood	0.8–0.9
Water	0.67
Coal	0.95
Fine snow	0.82
Glass	0.80

Table 7.1: *Different materials and their emissivity values.*

7.16 Properties of radiating thermal bodies

There are several very important properties of radiating bodies. Firstly, radiated power of an emitting body is given in its simplest mathematical form by an output power $P = \varepsilon \sigma A T^4$ watts, where ε is the emissivity, σ is the Stefan-Boltzmann constant (5.67×10^{-8} Wm^{-2} K^{-4}), A is the emitting surface area of the vessel or floating object, such as a man in the water, and T is the surface temperature of the ship or target in kelvin (K = °C + 273).

Consequently, radiated intensity 'I' is given by the equation:

$I = \varepsilon \sigma T^4$, where I is the power per unit area in Wm^{-2} (**eq 7.7**)

Example 7.6: If the emitting surface is human skin (emissivity = 0.98), what will be the radiated intensity at a temperature of 17°C (1 decimal place)?

Using: $I = \varepsilon \sigma T^4 = 0.98 \times 5.67 \times 10^{-8} \times (273.15 + 17)^4 = 393.8$ W

Another aspect of this equation is that the Intensity changes markedly for a small temperature change.

Example 7.7: If the object temperature is increased by 10 per cent, what will be the corresponding increase in radiated intensity (2 decimal places)?

Using the equation $I = \varepsilon \sigma T^4$

$I_T = \varepsilon \sigma T^4$ and

$I_{1.1T} = \varepsilon \sigma (1.1T)^4 = 1.46 \, \varepsilon \sigma T^4$

Thus $\dfrac{I_{1.1T}}{I_T} = 1.46$, a 46 per cent increase in radiated intensity for a 10 per cent increase in surface temperature.

The wavelength (λ_{peak}) at which the emission peak occurs is a function of absolute temperature T. This equation is known Wien's displacement law, named after Wilhelm Wien (1864–1928) in about 1893. This law shows the relationship between peak wavelength and radiated temperature.

$$\lambda_{peak} = \frac{2900}{T} \text{ microns} \qquad \textbf{(eq 7.8)}$$

Example 7.8: A tanker funnel has a uniform surface temperature of 57 °C. What is the peak emission wavelength (3 significant figures)?

$$\lambda_{peak} = \frac{2900}{(57 + 273.15)} = \lambda_{peak} = \frac{2900}{330.15} = 8.78 \text{ microns}$$

Curves of power per unit surface area emitted at each wavelength, plotted against wavelength for selected temperatures, show that hot bodies emit radiation over a relatively narrow range of frequencies. It has a well-defined peak wavelength, while cooler bodies emit radiation over a wider range of frequencies (figure 7.14, see plate section) with a less well-defined peak, much lower in radiated intensity.

7.17 Infrared transmission

Infrared interacts with matter differently when compared with light. For example, silicon and germanium are both opaque, with germanium appearing 'silvery' like a metal in the visible, but are both transparent to IR and are useful for thermal imaging lenses. In fact, germanium is more transparent in the IR than glass is in the visible, which reverses its reflective properties, being opaque in the infrared – hence the so-called greenhouse effect.

There is considerable infrared 'resonance' absorption due to gas molecules, particularly water vapour, found in Earth's atmosphere, so IR transmission at certain wavelengths over any useful distance is impractical (see Chapter 2). Knowledge of atmospheric IR transmission is vital for civilian and military applications, as vessel emission characteristics, sources and detectors cannot be separated from atmospheric propagation constraints.

Figure 7.15 (see plate section) shows the so-called atmospheric windows where transmission is good, and where it is poor due to absorption ('atmospheric holes'). Standard atmospheric transmission is shown for a 'clear' 3000 ft path of typical NW

European atmosphere. One observes essentially the same atmospheric transmission worldwide but it becomes more transparent in high mountainous and remote regions, such as the Himalayas, and a little lower in regions near sea level with high pollution level, e.g. the chemically induced smog surrounding major cities.

Design of a new sensing or communications system must take account of the atmospheric properties so it isn't designed at an inappropriate wavelength, e.g. 6.5 microns, where no sensor system works with useful range due to high transmission losses. Similar measures should be taken in the design of sonar; the underwater environment and its effect on acoustic waves must be considered as well.

The key atmospheric windows used are:

3–5 µm Detection systems, e.g. heat-seeking missiles, and short wave thermal imagers.

8–14 µm Thermal imagers and CO_2 lasers.

The carbon dioxide (CO_2) laser operates at 10.6 µm, well within the FIR window. This is the most important of the military lasers, with high power emission and long range. The eye lens is quite opaque (blocking the direct beam) and at this wavelength radiation presents a relatively low optical hazard. However, the Yttrium Aluminium Garnet (YAG) laser, used in tank range finders, operates at 1.06 µm, where the eye lens is transparent and presents a severe optical hazard.

At room temperatures, little radiated energy lies in the spectral band up to 5 µm; most appears in the longer wavelengths and is picked up by detectors sensitive in the 8–14 µm atmospheric window. 'Black body curves' in figure 7.14 show that room temperature objects at 300 K have their emission peak near 10 µm. These wavelengths transmit well in air.

7.18 Thermal Imaging

Thermal imaging is the 'viewing' of vessels by the heat they emit. IR cannot be seen by eye, so detected radiation must be converted to visible light to be seen. Older cameras used an objective lens to focus an image of a scene on to a cooled detector array via a rotating mirror.

Modern cameras use *staring arrays*, with a two dimensional pixel array used for imaging. A staring array or Focal Plane Array (FPA) is a rectangular array of light sensing pixels at the focal plane of a lens (figure 7.16). FPAs are commonly used for imaging (e.g. taking pictures or video imagery), but can be used for non-imaging purposes such as spectrometry and lidar.

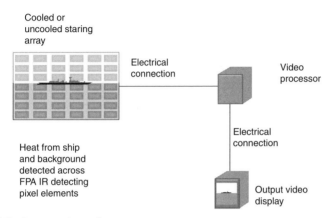

Figure 7.16: *Staring array thermal camera.*

Staring arrays are distinct from scanning array imagers in that they image the entire FOV at the same time without scanning. A staring array is like film in a camera; it directly captures a 2D image projected by the lens at the image plane, typically with 320 x 256 pixel element arrays. Arrays include InSb, QWIP, SLS and InGaAs detectors, which are affordable for non-military, non-scientific applications such as within some secondary schools. The most commonly used FPA types are InSb, InGaAs, HgCdTe (MCT) and Quantum Well (QWIP) devices. The latest use low cost, uncooled micro bolometers but are more expensive than visible counterparts.

7.19 Future infrared systems

There are already optical FPAs with 2048 × 2048 visible elements on integrated CCD wafers, with up to 100 million pixels per sensor for specialist camera applications. Transferring this to platinum silicide in the IR band is underway, with the intent to create an advanced prototype eyeball-size sensor with 2000 × 2000 pixels. Analysts anticipate multi-hyperspectral systems (imaging and simultaneously determining spectral content). Hyperspectral systems rely on diffraction gratings. Radiation is separated into different wavelength bands. Future systems will use Multispectral Image Fusion (I²/IR/CCD) and have less weight, smaller size, reduced power consumption, higher resolution, wider FOV and increased range for detection/ recognition and identification.

7.20 Thermal imagers and surveillance

Maritime surveillance thermal imagers can operate in total darkness and detect vessels through smoke or fog, night or day, but they have limitations – namely, image interpretation may be difficult at times.

At specific times of day, thermal contrast can disappear when the vessel radiated intensity and the radiated intensity of the sea are equivalent. Without some contrast, either with the intensity of the ship being greater than the background (the usual situation) or the ship intensity being less than the background, no contrast between ship and sea occurs. Heavy rain can cause uniform temperatures of all surfaces, resulting in loss of detail, the so-called thermal white-out.

Cameras using compressed gas cylinders must be replaced at regular intervals and these can only be used for relatively short periods of time. Thermal imagers cannot provide range information, but can show target detail as well as overall shape. It is now routine for military radar operators to identify contacts using thermal imagers. For range, a CO_2 laser range finder can be used.

Thermal imagers at sea have been developed with ranges over 20 km through the atmosphere. Actual range is determined by detector sensitivity, atmospheric attenuation and emitted target IR power. Range is largely a function of the viewing platform and viewing angle. Thermal cameras were used to observe the space shuttle on re-entry over 3000 km away and of course may be pointed towards stars to detect heat from objects light years away!

7.21 Typical ship's IR emissions

From an imaging perspective, ship IR emissions are divided into Infra Red Cross Section (IRCS), the total IR power emitted by the maritime platform, and Infra Red Signature (IRS), the detailed distribution of IR emitters on the vessel.

In the 3–5 micron band, hot objects – ship's propulsion units, exhaust plume, ship's uptake (a surface for drawing up air) and funnel surfaces – dominate. The MIR band gives high contrast but poor edge definition, such as looking at a bright torch shining on a dark night. It is easy to detect, but what is it? MIR band imaging sensors are in development but most imaging takes place in the FIR band.

The 8–14 micron band covers 'room temperature' surfaces and allows small temperature differences to be detected with great sensitivity, giving high definition pictures in the FIR band. It is in this region that surveillance thermal imagers can examine a ship's IR Signature.

Identification in this band comes mainly from strong contrast edges due to differential heating across the ship. Machinery spaces and refrigerated compartments are observed to provide well-defined hot and cold spots on a ship's hull.

For military vessels, stealth has become of great interest recently. Stealth is the ability for the vessel to be undetected in particular sensor bands. The advantage of not being detected should be obvious, but greater discussion of stealth is found elsewhere [7.9]. There are practical things naval ships will incorporate, such as the use of low emissivity coatings and alteration of IR signature by cooling the plume and funnel surfaces to reduce overall heat output. This benefits civilian shipping vessels as well, as reduced carbon footprint saves money and improves environmental management. However, merchant shipping is often more critical on saving time, which doesn't necessarily equate to reduced emissions.

IR emission is a surface phenomenon; coating ships in low emissivity paint reduces radiated IR, reducing IRCS. Exhaust plumes are the biggest contributor to IRCS between 300–500 °C. Although gases are poor IR radiators, a vessel's plume generally contains hot carbon particles that radiate almost perfectly. Naval engine exhausts can be physically separated from a funnel's outer casing, reducing the temperature reached by the funnel surface and carbon 'scrubbed' from emissions.

7.22 Thermal contrast

Radiation emitted by a body depends on temperature and on its surface emissivity. For good imaging, contrast between target radiation and that from the background must be high. For all detection, contrast is the fundamental issue, as detection relies on a detectable difference observed between that detected and its background.

Let us compare radiated intensities between two bodies and see if there is sufficient contrast between them to be separated by an imager.

Radiated intensity is given by: $I = \varepsilon \sigma T^4$. For two different objects, '1' and '2', to be distinguished, $\varepsilon_1 \sigma T_1^4$ must be different from $\varepsilon_2 \sigma T_2^4$. The condition where they are equal is called thermal crossover, $(\varepsilon_1 \sigma T_1^4 = \varepsilon_2 \sigma T_2^4)$.

So that $\varepsilon_1 T_1^4 = \varepsilon_2 T_2^4$. The operating environment objects cannot be distinguished at crossover.

Example 7.9: What is the average Kelvin temperature of a bridge if a river at 4 °C has an emissivity value of 0.92 and cannot be distinguished against a bridge with an emissivity of 0.85 (1 decimal place)?

Using the relationship:

$$\varepsilon_b T_b^4 = \varepsilon_r T_r^4$$

Rearranging:

$$T_b^4 = (\varepsilon_r T_r^4) / \varepsilon_b$$

and substituting:

$$T_b^4 = \frac{0.92 \times (273.15 + 4)^4}{.85}$$

$$\text{So } T_b = \sqrt[4]{\frac{0.92 \times (273.15 + 4)^4}{.85}} = 282.7 \, K$$

In summary, and with the various sensor systems considered so far, let us compare and evaluate their capabilities and limitations.

SENSOR SYSTEM	Typical detection range at sea level	Actual range and resolution	Passive or active	Operability in presence of light rain	Day or night effects
Radar	Early warning 400 km +	Yes	Active	Good	Little change
Image Intensifier	Up to 20 km	No	Passive	Bad	Generally used during night or dawn/dusk
Thermal Imaging Camera	Up to 30 km	No	Passive	Not as good	Big changes at dawn/dusk at thermal crossover
Laser	Up to 100 km	Yes	Active	Strongly attenuated	None

Table 7.2: *Comparison between different sensors.*

Selection of a particular sensor depends upon its capabilities and limitations, and is modified by operating requirements – the need for day/night capability, operating altitude/speed and maritime factors: imaging FOV requirements, image resolution, and the requirement for real-time information, etc. Cost should include development, purchasing, integration, maintenance and support, etc., especially in a harsh maritime environment. Platform limitations should be considered, e.g. are there size and weight constraints, power requirements, data storage and transfer issues?

No single sensor or communications device fulfils all the design and operational requirements. It is common to fit several different sensors (radars or imagers) to a platform to provide the required platform capabilities.

The capabilities of different systems are not identical, and are affected differently by weather. Hence, different sensors on different present/future platforms are important to ensure successful environmental exploitation.

7.23 Operating environment

The environment in which maritime platforms operate has a significant impact upon platform design. Sea, land and air platforms operating at low level will operate in a rapidly fluctuating and changing environment. Environmental changes significantly affect maritime operations. Sensors can experience high thermal noise and clutter backgrounds. Naval crew often operate in poor visibility conditions due to battlefield obscuration, as well as the common problems experienced by civilian and military crew in dust and sand conditions of the Gulf, contrasted against snow conditions in the Atlantic. Overall, maritime platforms, whether manned (e.g. ships and oil rigs) or unmanned, contend with harsh operating conditions on a daily and annual basis, the most dangerous working environment. In addition, salt spray generates a highly corrosive atmosphere. These conditions also affect communications.

The best way to utilise a platform's sensors is by integrating the different sensor outputs, creating hybrid and complex displays. Utilisation of platform sensors is central to the naval concept of Network Enabled Capability (NEC), requiring digitisation of the battlespace for handling large volumes of data. Data fusion and false colour imagery is becoming increasingly common. The success of 21st-century civilian maritime safety and military engagement will rely heavily on successful integration of different sensing and communications systems.

Self-assessment Questions

7.1   Explain photoelectric and secondary electron emission.

If an incident photon frequency is 750 nm, and the work function of the electron emitting surface is 1.8 eV, what is the maximum Kinetic Energy of the emitted electrons in eV (2 decimal places)?

7.2   Explain how electron effects are used in a practical MCP image intensifier.

If 14 electrons are emitted on average for every 5 incident electrons, what is the secondary electron emission coefficient?

7.3   Explain how a simple laser operates. Briefly describe the different applications of lasers in the marine environment.

7.4   If the initial energy state of the excited atoms in the laser medium is 1.75 eV and the final energy state of the laser medium is 2.9 eV, what is the frequency of emitted laser radiation?

7.5   State what is meant by NIR, MIR and FIR radiation, and where the two main atmospheric transmission windows occur. Why are they important for sensing?

7.6   Explain how a low emissivity coating can reduce the IRCS of a ship. If the emitting surface has an emissivity of 0.27, what is the radiated intensity at 17 °C (1 decimal place)?

7.7   State the important laws and properties of infrared radiation.

If a black body emits 100 W and a test material emits 92 W, what is the material emissivity (2 decimal places)?

An aircraft exhaust funnel has a uniform surface temperature of 250 °C. What will be the peak emission wavelength observed (2 significant figures)?

7.8   State the advantages and disadvantages of thermal imaging compared with radar. If the object temperature is increased by 15 per cent, what will be the corresponding increase in the radiated intensity (2 decimal places)?

7.9   Explain what is meant by thermal contrast and thermal crossover. What is the average Kelvin temperature of a bridge if a river at 4 °C has an emissivity value of 0.92 and cannot be distinguished against a bridge with an emissivity of 0.85 (1 decimal place)?

7.10   Discuss techniques that can be used to reduce a ship's IRS.

Assuming an inverse square law fall-off in intensity with distance, suggest an expression to give the thermal intensity at one distance in terms of the thermal intensity at a different distance.

Now use your expression to evaluate the relative change in intensity if the new emissivity value is half the original value, and the new temperature is twice the original value.

REFERENCES

[7.1] 'Long Lifetime Generation IV Image Intensifiers with Unfilmed Microchannel Plate', JP Estrera et al, Proc. SPIE Vol. 4128. Image Intensifier and Applications II (2000).

[7.2] www.gov.uk/government/publications/laser-radiation-safety-advice/laser-radiation-safety-advice#fn:2

[7.3] 'Stimulated Optical Radiation in Ruby', TH Maiman, *Nature* 187 (6 August 1960), pp. 493–494. *Bibcode:1960Natur.187..493M. doi:10.1038/187493a0.*]

[7.4] 'Optical and Microwave-Optical Experiments in Ruby', TH Maiman, *Physical Review Letters*, Vol. 4 (11) (1 June 1960), pp. 564–566. *Bibcode:1960PhRvL...4..564M. doi:10.1103/physrevlett.4.564.]*

[7.5] 'Repetitively Pulsed Tunable Dye Laser for High Resolution Spectroscopy', TW Hänsch, *Applied Optics*, Vol. 11 (1972), pp. 895–898.

[7.6] 'Neodymium-doped silica single-mode fibre lasers', RJ Mears, L Reekie, SB Poole, DN Payne, *Electronics Letters* Vol. 21 (17) (15 August 1985), pp. 738–740.

[7.7] 'Blue-Green Dye Lasers for Underwater Illumination', TG Pavlopoulos, *Naval Engineers Journal*, Volume 114, Issue 4 (October 2002), pp. 31–40.

[7.8] *Night Vision: Exploring the Infrared Universe*, M Rowan-Robinson (Cambridge University Press, 2013, 978–1-1070–2476–2), p.23.

[7.9] *Stealth Warship Technology*, C Lavers (Reeds Marine Engineering and Technology Series, Volume 14, 2012, ISBN 978–1-4081–7552–1).

8

Common Maritime System Monitoring Sensors and Transducers

'There is no safe amount of radiation. Even small amounts do harm.'   Linus Pauling PhD, Nobel Laureate

8.1 Sensor types covered

A broad range of sensors are found on board ships today, including: temperature and pressure, radioactivity monitoring, speed transducers, flow, force, displacement, accelerometers, range, depth and strain gauges, among others. We will examine a few key operating principles, focusing on pressure, temperature and resistance. Most modern analogue equipment works approximately in the current range 4–20 mA, pressure systems are typically based in the range 0.2–1.0 bar, and temperature 0–1500 °C, but sensors operate outside of these ranges as well.

There are many things to measure at sea besides temperature, pressure and current: flow rate, force, displacement, velocity, acceleration, strain and stress, mass, depth (level), size (volume), salinity and pH (alkalinity/acidity). Such sensors can mostly be operated using simple on/off switches, or a thermostat (heat), or level, pressure or proximity switches, etc. There is a wide range of commercial sensors available for each industrial maritime application, including resistance (Pt100/Pt1000), thermocouples (NiCr/NiAl), thermistor elements, as well as wireless temperature sensors based on Surface Acoustic Wave (SAW) devices.

But perhaps we should first ask the question: 'What exactly *is* a sensor or transducer?' A sensor is any device that provides a useful output in response to a specific quantity or *measurand*. A transducer is the active element of a sensor, and is regarded as the *sensor head* – distinct from any associated circuitry.

We have introduced input and output transducers in Chapter 1 (1.3.1). However, the general principle applies equally to sensors, where there are many possible input and output devices. Transducers will be discussed in terms of particular sensing principles rather than applications, as each potential sensing principle can have several applications. Some systems can be used together, e.g. Venturi with manometer. When the input is a physical quantity and the output electrical, we call the transducer a sensor, but if the input is electrical and the output a physical quantity, e.g. microphone, we better describe this as an actuator. Commonly measured quantities are given in table 8.1.

Measurand	Sensor/Transducer
Position, velocity, acceleration, force, strain, stress, pressure, torque	Mechanical
Temperature, flux, specific heat, thermal conductivity	Thermal
Magnetic field, flux, permeability	Magnetic
Charge, voltage, current, electric field, conductivity, permittivity	Electrical
Gas and liquid fluid concentration	Chemical and biological
Wave velocity, spectrum, amplitude	Acoustic
Refractive index, reflectivity, absorption	Optical

Table 8.1: *Common measurands and sensor types.*

8.2 Temperature transducers

8.2.1 Thermoelectric sensor

Temperature transducers frequently use a thermocouple. When two wires with different electrical properties are joined at one end and one junction made hot and the other cold, a small electrical current is produced proportional to the temperature difference. This phenomenon was discovered by Thomas Seebeck (1770–1831) in 1821, with the hot junction forming the sensor head. He found that junctions of dissimilar metals produce an electric current when exposed to a temperature gradient. Thermoelectric materials can be used for harvesting heat and refrigeration and assembled into mechanical structures that transform heat into electrical energy. Thermoelectric materials are used in specialist cooling applications, e.g. maintaining a stable temperature in lasers and optical detectors, and are found in office water coolers. They are used in space exploration to convert heat from radioactive materials into electricity. Focus on energy sustainability and stricter legislation on CO_2 emissions imposed on automobile manufacturers has increased interest in these materials. About one-third of an internal combustion engine's fuel energy is converted into mechanical energy, with the remainder lost as heat. A thermoelectric generator 'harvests' exhaust waste heat gases, which at 300–500 °C can turn this into electricity. State-of-the-art modules generate about 1 kW, which can be used to power vessel electrical equipment. In a car this allows for a smaller alternator, reducing roll friction, leading to increased fuel efficiency and reduced CO_2 emissions. The Seebeck effect is the physical basis for the thermocouple, often used for temperature measurement according to equation (8.1).

$$V = a(T_h - T_c) \qquad \textbf{(eq 8.1)}$$

A voltage difference, V, is produced across the terminals of an open circuit made from dissimilar metals, A and B, whose two junctions are held at different temperatures. The voltage is directly proportional to the difference between the hot and cold junction temperatures, $T_h - T_c$. The voltage or current produced across the junction is caused by diffusion of electrons from a high electron density region to a low electron density region, as the electron density is dissimilar in different metals. Conventional current flows in the opposite direction to electron flow.

The Peltier effect, named after the French physicist Jean Peltier (1785–1845), is the presence of heating or cooling at an electrified junction of two different conductors (figure 8.1). When a current is made to flow through a junction between two conductors A and B, heat may be generated (or removed) at the junction. The Peltier heat generated at the junction per unit time, $\dfrac{dQ}{dT}$, is equal to:

$$\frac{dQ}{dT} = (\Pi_A - \Pi_B)I \qquad \textbf{(eq 8.2)}$$

where ω_A and (ω_B) are the Peltier coefficients of conductor A and B, and I the electric current (from A to B).

Most thermocouple metals have a relationship between the two temperatures and the Electro Motive Force (e.m.f.) or ε, which is given by:

$$\varepsilon = a(T_h - T_c) + \beta\,(T_h - T_c)^2 \qquad \textbf{(eq 8.3)}$$

where a and β are constants for the particular thermocouple. The relationship is nearly always linear over the operating range.

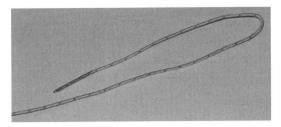

Figure 8.1: *A typical thermocouple.*

The actual characteristic and suitable operating temperatures depend upon the metals used in the wires. It is important that thermocouples are standard so the same e.m.f. ε represents the same temperature.

Thermocouples come in many forms: wires insulated from each other with plastic or glass fibre materials (figure 8.1). For high temperature work, wire pairs are placed within an insulated tube. For industrial uses, sensors come in a metal enclosure such as stainless steel.

The response time of a thermocouple is a very important factor, and can be expressed by a time constant, T. A temperature sensor at temperature T_1 is immersed in a liquid at temperature T_2. T is the time it takes before the temperature sensor reaches $1 - 1/e$ of the temperature difference $T_2 - T_1$, approximately 63 per cent of its final value. After time 2.3T, 90 per cent of the temperature difference is reached, etc. If the sensor is mounted in a protected well, the time constant will be approximately double.

8.2.2 Resistance type sensors

These devices rely on changes in electrical resistance in the conductor due to a temperature change. If a constant voltage is applied to the conductor, a current flows through it, with current varying as temperature varies because conductor resistivity changes with temperature.

Most conductors' resistance changes when their temperature changes. Usually this change follows a straight line relation. For pure metals, resistance increases continuously with temperature. For certain metal alloys used for making resistors, e.g. Manganin or constantan, resistance is largely unaffected by temperature. For other partial conductors, e.g. carbon, resistance decreases with temperature.

It is possible to deduce a straight line relationship between resistance and temperature for resistors.

$$R = R_0(1 + aT)$$ **(eq 8.4)**

where T is the temperature in centigrade and a is the temperature coefficient of the conductor in question. Copper and platinum are often used. Copper has $a = 0.004265$ per degree centigrade while platinum, which is often used, has $a = 0.00385$ per degree.

A basic temperature sensor is made by winding a thin resistance wire into a small sensor head. Wire resistance represents the temperature. This has an advantage over a thermocouple in that it is unaffected by the temperature of the equipment or gauge end. A typical operating range is −200 to 400 °C.

Example 8.1: The cold resistance of a coil of wire is 20 Ω at 15 °C. It is heated to give a resistance of 23 Ω. Find its temperature rise, if the temperature coefficient of the resistance material is 0.0042 per degree centigrade (1 decimal place).

Thus, using $R = R_0(1 + aT)$ (eq 8.4)

for the two temperatures T_1 and T_2:

$$\frac{R_2}{R_1} = \frac{R_0(1 + aT_2)}{R_0(1 + aT_1)}$$ **(eq 8.5)**

Thus:

$$\frac{23}{20} = \frac{R_0(1 + 0.0042T_2)}{R_0(1 + 0.0042T_1)}$$

Hence: $1.15 \times 1.063 = 1 + 0.0042T_2$

So $T_2 = 53.1\ °C$

And the temperature rise is $53.1 - 15 = 38.1°C$

8.2.3 Potentiometer

A potentiometer has a variable electrical resistance. A length of resistance material has a voltage applied over its ends. A slider moves along it (either a linear or rotary pick up), yielding the voltage at its position or angle. Tracks may be made from carbon, resistance wire or a piezoresistive material. The wire wound type produces small step changes in output, depending on how fine the wire is and how closely it is coiled around the track.

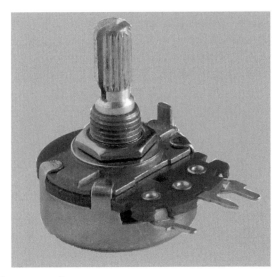

Figure 8.2: *A single turn potentiometer.*

8.2.4 The thermistor

An unusual type of resistance sensor for measuring temperature is the thermistor (thermal resistor), made from a small piece of semiconductor material. The material is unusual because the resistance changes a lot for a small temperature change, so a smaller sensor head is required and costs less than platinum wire. However, the thermistor temperature range is generally limited so different thermistor values are required for different temperature ranges, typically −90 to 130 °C [8.1], and are often used in small handheld thermometers for everyday practical applications. The relationship between resistance and temperature is of the form:

$$R = Ae^{\frac{B}{T}} \qquad \textbf{(eq 8.6)}$$

Example 8.2: The characteristic of a thermistor or Resistance Temperature Device (RTD) in terms of its resistance vs temperature generally follows the absolute value of its temperature described by the following equation:

$$\frac{R_{T_1}}{R_{T_2}} = e^{\beta\left(\frac{1}{T_1} - \frac{1}{T_2}\right)} \qquad \textbf{(eq 8.7)}$$

$$\text{where: } \beta = \frac{\ln\left(\frac{R_{T_1}}{R_{T_2}}\right)}{\left(\frac{1}{T_1} - \frac{1}{T_2}\right)} \text{ and}$$

R_{T_1} is the zero power resistance at absolute temperature T_1,

R_{T_2} is the zero power resistance at absolute temperature T_2, and

β is a constant whose value is determined by the thermistor material.

Find the β of a thermistor at $R_{30/50}$ whose values are $R_{30} = 9000$ ohms and $R_{50} = 3200$ ohms (4 significant figures).

$T_1 = 273.15 + 30 = 303.15$ K $T2 = 273.15 + 50 = 323.15$ K

$$\beta = \frac{\ln\left(\frac{9000}{3200}\right)}{\left(\frac{1}{303.15} - \frac{1}{323.15}\right)}$$

$\beta = 5903$

8.2.5 Bimetallic types

If two metals are rigidly joined together as a two-layer strip and heated, the difference in the metals' expansion rates will cause the strip to bend. This effect can convert temperature changes into mechanical displacement. The strip consists of two strips of dissimilar metals, which, when heated, expand at different rates. The materials are often a combination of steel and copper, or steel and brass. Strips are joined together by brazing, welding or riveting. The different expansions force a flat strip to bend one way if heated, and in the opposite direction if cooled. The metal with the higher coefficient of thermal expansion is on the outside of the curve when the strip is heated, and on the inside when cooled. John Harrison, an 18th-century English clockmaker, is credited with the invention of the first surviving bimetallic strip made for his third marine chronometer of 1759 to compensate for temperature-induced changes in the balance spring [8.2].

They can be made to operate a thermostat (e.g. in a boiler), in limit switches, or to set off alarms. In the regulation of heating and cooling, bimetallic thermostats operate over a wide range of temperatures, making or breaking contact due to the deflection. In these, one end of the bimetallic strip is mechanically fixed and attached to an electrical power source, while the other, freely moving end carries an electrical contact. In the industrial type, the strip is twisted into a long thin coil inside a tube. One end is fixed at the bottom of the tube and the other turns and moves a pointer on a dial.

If unbonded, the free lengths of each material would be different after a temperature change (figure 8.3). When bonded, the difference in unconstrained lengths gives rise to internal stresses within the strip, causing it to bend.

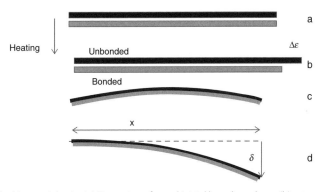

Figure 8.3: *The bimaterial strip: (a) Two strips of equal initial length undergo (b) a temperature change ΔT, the relative difference in unconstrained lengths is Δε. (c) Since the two strips are bonded together, the resulting internal stresses generate a uniform curvature. (d) The clamped strip allows deflection measurement related to the curvature.*

When thermal equilibrium is reached, the resulting curvature, κ (reciprocal of the radius of curvature), is related to displacement, δ, and the distance, x, along the strip at which the displacement is being measured by the relationship:

$$K = \frac{2\sin[\tan^{-1}(\frac{\delta}{x})]}{\sqrt{(x^2 + d^2)}} \qquad \textbf{(eq 8.8)}$$

Curvature is also related to the metals' properties and dimensions through the equation:

$$K = \frac{6E_A E_B(h_A + h_B)h_A h_B \Delta \in}{E_A^2 h_A^4 + 4E_A E_B\, h_A^3 h_B + 6E_A E_B\, h_A^2 h_B^2 + 4E_A E_B\, h_A\, h_B^3 + 4E_A E_B\, h_B^3 + E_B^2 h_B^4} \qquad \textbf{(eq 8.9)}$$

where E_A, E_B are the relevant Young's moduli, h_A and h_B are the thicknesses of the two materials A and B. The 'mismatched' strain, $\Delta\varepsilon$, is given by:

$$\Delta\varepsilon = (a_A - a_B)\Delta T \qquad \textbf{(eq 8.10)}$$

where a_A and a_B are the thermal expansivities of the two materials.

It can be seen that the curvature depends not just on the expansivity mismatch and temperature change but also on the relative stiffness and thicknesses of the two materials. If the two strips are of equal thickness (h), and the stiffness ratio is termed E_*, the equation can be rewritten:

$$\Delta\varepsilon = \frac{Kh(E^* + 14 + \frac{1}{E^*})}{12} \qquad \textbf{(eq 8.11)}$$

which means the curvature increases as the strips are thinner. Curvature is also smaller if one of the materials has a much greater stiffness than the other.

It is possible to estimate the value of ΔT corresponding to a measured curvature of the steel-aluminium bimetallic strip and even to estimate the temperature of boiling nitrogen.

Example 8.3: For the steel-aluminum bimetallic strip, the following measurements were made: $\delta = 0.021$ m, $x = 0.190$ m, $h = 1$ mm, the initial temperature is 22 °C:
$E^* = 21070$ $a_A = 1.5 \times 10^{-5}$ K^{-1} $a_B = 2.3 \times 10^{-5}$ K^{-1}. What is the temperature of the boiling nitrogen (1 decimal place)?

Using: $\kappa = \dfrac{2\sin\left[tan^{-1}\left(\frac{\delta}{x}\right)\right]}{\sqrt{(x^2 + d^2)}}$

$K = \dfrac{2\sin\left[tan^{-1}\left(\frac{0.021}{0.190}\right)\right]}{\sqrt{(0.021^2 + 0.190^2)}} = 1.1493939$

From the equation for the mismatched strain:

$\Delta\varepsilon = \dfrac{Kh\left(E^* + 14 + \frac{1}{E^*}\right)}{12}$

$\Delta\varepsilon = \dfrac{1.1493939 \times 0.001\left(210/70 + 14 + \frac{1}{210/70}\right)}{12}$

$= 1.6602 \times 10^{-3}$

And from $\Delta\varepsilon = (a_A - a_B)\Delta T$

$\Delta T = \dfrac{\Delta\varepsilon}{(a_A - a_B)} = \dfrac{1.6602 \times 10^{-3}}{(1.5 \times 10^{-5} - 2.3 \times 10^{-5})} = -207.5$ Kelvin

T boiling $= 22 - 207.5 = -185.5$ K (which is incorrect!)
Compare this value with the accepted figure of -196 °C.

8.3 Liquid expansion glass thermometers and vapour pressure sensors

These are thermometers filled with a liquid such as mercury or an evaporating fluid such as used in refrigerators. Mercury is used for hot temperatures and coloured alcohol for cold temperatures. In both cases, the inside of the sensor head and the connecting tube are completely full. Any rise in temperature produces expansion or evaporation of the liquid so the sensor becomes pressurised. Pressure is related to the temperature and may be indicated on a simple pressure gauge.

Pressure may be converted into an electrical signal or directly operate a thermostat. These instruments are robust and used over a wide range and can be fitted with electric switches to set off alarms. The problems with glass thermometers are that they are brittle, while mercury solidifies at −40 °C while alcohol boils at around 120 °C. Accurate manufacture is needed and this makes accurate thermometers expensive. It is also easy for people to make mistakes reading them, especially outside in the dark and cold. Glass thermometers are not generally used today in industry but, if they are, they are protected by a shield from accidental breakage.

8.4 Aneroid and manometer pressure transducers

Pressure sensors convert the pressure into either mechanical movement or an electrical output. Complete gauges sense the pressure and then indicate them on a dial or scale.

8.4.1 Aneroid gauge

Aneroid gauges have a metallic pressure sensing element that 'flexes' or moves mechanically under the effect of a pressure difference across the element. 'Aneroid' means 'without fluid', and the term distinguishes these gauges from hydrostatic gauges. However, aneroid gauges can measure the pressure of a liquid as well as a gas, and they are not the only type of gauge that operates without fluid. For this reason, they are often called mechanical gauges. Aneroid gauges are not dependent on gas being measured, and are less likely to contaminate the system than hydrostatic gauges.

The pressure sensing element may be a Bourdon tube, a diaphragm, a capsule, or a set of 'bellows' that changes shape in response to the pressure of the region. Deflection of the pressure sensing element is often read by a linkage connected to a needle, or read by a secondary transducer. The most common secondary transducer in a modern vacuum gauge measures a change in capacitance due to the mechanical deflection. Gauges that rely on a change in capacitance are often referred to as capacitance manometers.

8.4.2 Liquid pressure column difference

By considering pressure, liquids can be used in instrumentation where gravity is present. Liquid column gauges consist of a vertical column of some liquid in a tube that has its ends exposed to different pressures. The column rises or falls until its weight (force due to gravity) is in equilibrium with the pressure difference between the two ends of the tube (force due to fluid pressure). The simplest version is a U-shaped tube about half full of liquid, one side connected to the region of interest while the reference pressure (e.g. the atmospheric pressure or a vacuum) is applied to the other (figure 8.4). The difference in liquid level represents the applied pressure. The pressure exerted by a fluid column of height h and density ρ is given by the hydrostatic pressure equation: $P = hg\rho$. The pressure difference between applied pressure Pa and the reference pressure P_0 in a U-tube manometer is found by solving $Pa - P_0 = hg\rho$. The pressure on either end of the liquid must be balanced (since the liquid is static) and so $Pa = P_0 + hg\rho$ (**eq 8.12**) (figure 8.4).

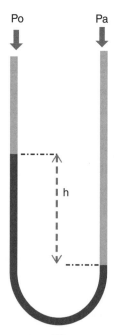

Figure 8.4: *Manometer liquid column pressure difference.*

In most liquid column measurements, the result of the measurement is the height, h, expressed typically in mm. The 'h' is known as the *pressure head* specified in units of length and the measurement fluid specified. When accuracy is critical, the measurement fluid temperature is also specified, because liquid density is a function of temperature. For example, the pressure head might be written '752.2 mm Hg at 55 °C' for measurements taken with mercury or water as the manometric fluid, respectively. The word 'gauge' or 'vacuum' may be added to distinguish between pressure above or below atmospheric pressure. Both mm of mercury and inches of water are common pressure heads, which can be converted to SI units of pressure.

Although any fluid can be used, mercury is preferred for its high density (13.534 g cm^{-3}) and low vapour pressure. For low pressure differences, light oil or water are used. Liquid column pressure gauges have a highly linear calibration. They have poor dynamic response because the fluid in the column may react slowly to a pressure change. When measuring vacuum, the working liquid may evaporate and contaminate the vacuum if its vapour pressure is too high. When measuring liquid pressure, a loop filled with gas or a light fluid can isolate the liquids to prevent them from mixing – e.g. mercury is used as the manometer fluid to measure differential pressure of a fluid such as water. Simple hydrostatic gauges can measure pressures ranging from a few torr (a few 100 Pa) to a few atmospheres.

A single-limb liquid column manometer has a larger reservoir instead of one side of the U-tube and a scale beside the narrower column. The column may be inclined to amplify liquid movement. Based on the use and structure, the following manometer types are available: simple manometer, micromanometer, differential manometer and inverted differential manometer. Hydrostatic pressure sensors can include hydrostatic gauges (such as the mercury column manometer), piston-type gauges and liquid column systems, which have already been discussed.

Mechanical movement is produced with the following elements:

Bourdon tube A hollow tube with an elliptical cross section. When a pressure difference exists between the inside and outside, the tube tends to straighten out and the end moves. The movement is usually coupled to a needle on a dial to make a complete gauge.

Spring and piston The pressure acts directly on the piston and compresses the spring. The position of the piston directly relates to the pressure. A window in the outer case allows pressure to be indicated. This type is usually used in hydraulics, where the ability to withstand shock, vibration and sudden pressure changes is needed.

Bellows and capsules These are hollow, flattened structures made from thin metal plate and can be made of several capsules. When pressurised, the bellows expand and produce mechanical movement. If the bellows is encapsulated within an outer container, the movement is proportional to the difference between the pressure on the inside and outside. Bellows and single capsules are used in many instruments and are useful for measuring small pressures. The bellows configuration is used in aneroid barometers (barometers with an indicating needle and dial card), altimeters, altitude recording barographs, and altitude telemetry instruments used in weather balloon radiosondes.

Diaphragms Similar to the capsule, but the diaphragm is very thin and often made of rubber. The diaphragm expands when very small pressures are applied. The movement is transmitted to a pointer on a dial through a fine mechanical linkage such as in an aneroid barograph (figure 8.5).

Figure 8.5: *Barograph.*

8.5 Electrical pressure transducers

There are many ways of converting mechanical movement of the preceding types into electric signals. The most common types that directly produce an electrical signal are listed below.

8.5.1 Piezoelectric types

The element is a piece of crystalline material, often quartz, that produces an electric charge on its surface when mechanically stressed. The electric charge may be converted into voltage. This principle is used in the 'pick-up' crystal of a record player, in microphones and even to generate a spark as in a domestic gas igniter. When placed inside a pressure transducer, pressure is converted into an electric signal.

8.5.2 Strain gauge types

Strain gauges are small elements fixed to a surface that is strained (see section 8.8.2). The change in length of the element produces changes in electrical resistance, which is converted into a voltage. A typical pressure transducer contains a metal diaphragm, which bends under pressure.

8.5.3 Other pressure transducer types

Other electrical pressure sensors include optical (e.g. a physical change of an optical fibre to detect strain due to applied pressure), potentiometric, resonant frequency shift and capacitive systems. Magnetic systems, measuring the displacement of a diaphragm from a change in inductance, often with the Hall effect or via the eddy current principle, are possible.

8.6 Typical optical sensors

Common optical systems here include lasers for replenishment at sea and range finding. A few position sensors operate using laser Doppler vibrometers or rotary optical systems. Light sensors are used in cameras, infrared detectors and lighting applications. Sensors are usually composed of a photoconductor such as a photoresistor, photodiode or phototransistor.

8.6.1 Optical position sensors

One common example is found on machine tools, where they measure the position of the work table and display it in digits on the gauge head. The basic principle is as follows: light is emitted through a transparent strip or disc into a photoelectric cell. The strip or disc has fine lines engraved on it, which interrupt the beam. The number of interruptions is counted electronically and this is represented on a display as a position or angle. Photogates are used in counting applications with a transmitter and receiver at opposite ends of the sensor; the time at which light is broken is recorded. Ultrasonic sensors are used for position measurements in the range 2–13 MHz up to 15–20 degrees either side of the direct output path (figure 8.6).

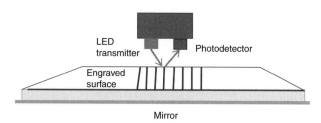

Figure 8.6: *Light source and photodetector positional sensor.*

8.6.2 Proximity sensor

A proximity sensor detects nearby objects without physical contact, and often emits an electromagnetic field or a beam of radiation (infrared, for instance), and looks for changes in the field or return signal. The object being sensed is often referred to as the proximity sensor's target. With magnetic field sensors, e.g. sensors used for power steering and current measurements on transmission lines, the Hall voltage induced is found to be proportional to the magnetic field.

8.6.3 Photoresistor

Resistance depends on the intensity of light incident upon it. Under dark conditions, resistance is quite high (mega ohms) and is called dark resistance. Under bright light conditions, resistance is lower (just a few hundred ohms). It takes a few milliseconds to respond to bright light, but several seconds to return to its original dark state. Photoresistors exhibit a linear response for optical illuminations versus resistance.

There are also many new optical sensors being developed currently.

8.7 Flow meters

There are many types of flow meters depending on the application, classified as follows: position displacement, inferential, variable area and differential pressure types.

8.7.1 Positive displacement types

These have a mechanical element that makes the shaft of the meter rotate once for an exact known quantity of fluid. The quantity of fluid depends on the number of revolutions of the meter shaft and the flow rate depends upon the speed of rotation. Both the revolutions and speed may be measured with mechanical or electronic devices. Some common types are: rotary piston and vane types, lobe type or meshing rotor, reciprocating piston type and fluted spiral gear.

8.7.2 Inferential type meters

The flow of fluid is 'inferred' from some effect produced by the fluid flow. Usually, this is a rotor that is made to spin and the speed of the rotor is sensed mechanically or electronically. The main types are: turbine rotor, rotary shunt, rotary vane and helical turbine.

8.7.3 Variable area types

There are two very common types of this meter: float type (rotameter) and tapered plug.

Float type The float is inside a tapered tube (figure 8.7).

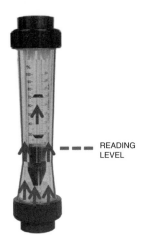

READING LEVEL

Figure 8.7: *Float type.*

Fluid flows through the annular gap around the edge of the float. The restriction causes a pressure drop over the float and the pressure forces the float up. Because the tube is tapered, the restriction is decreased as the float moves up. Eventually, a level is reached where the restriction just produces a pressure force, which counteracts the float weight. The float level indicates the flow rate. If flow changes, the float moves up or down to find a new balance position. When dangerous fluids are used, protection is needed against tube fracturing, which may be made of a non-magnetic metal. The float has a magnet on it. As it moves up and down, the magnet moves a follower and pointer on the outside. The position of the float may be measured electrically by building a movement transducer into the float.

Tapered plug type In this meter, a tapered plug is aligned inside a hole or orifice. A spring holds it in place. Flow is restricted as it passes through the gap and a force is produced, which moves the plug. Because it is tapered, the restriction changes and the plug takes up a position where the pressure force just balances the spring force. Plug movement is transmitted with a magnet to an indicator on the outside.

8.7.4 Differential pressure flow meters

These are a range of meters that convert flow rate into a differential pressure. The important types are: orifice, Venturi, nozzle meters and pitot tubes.

Orifice meter An orifice plate (figures 8.8a and 8.8b) is simply a plate with a hole through it, placed in the flow path, which constricts flow. Measuring the pressure differential across the constriction gives the flow rate. It is a basic type of Venturi meter, but with higher loss.

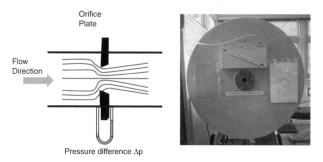

Figure 8.8 a: *Orifice plate.* 8.8 b: *Another orifice plate.*

Venturi meter A Venturi meter (figure 8.9a) constricts the flow in some fashion, and pressure sensors measure the differential pressure before and within the constriction (figure 8.9b). This method is widely used to measure flow rate in the transmission of gas through pipelines, and has been used since antiquity. The coefficient of discharge of a Venturi meter ranges from 0.93 to 0.97.

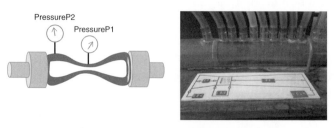

Figure 8.9a: *Venturi meter.* 8.9b: *Venturi meter with attached manometer.*

Pitot tubes These are pressure measuring instruments, used to measure fluid flow velocity by determining the stagnation and static pressures. Bernoulli's equation is used to calculate the dynamic pressure and hence fluid velocity.

Let a small volume Δ flow through the tube in the direction shown in figure 8.10.

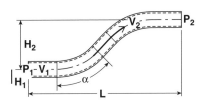

Figure 8.10: *Small volume flow.*

Work done on the system $= p_1\Delta$

Work done *by* the system $= p_2\Delta$

So net work done on a system $= p_1\Delta - p_2\Delta$

Corresponding mass of a volume Δ is $\rho\Delta$ where ρ is the density

Gain in kinetic energy $= \dfrac{1}{2}\rho\Delta v_2{}^2 - \dfrac{1}{2}\rho\Delta v_1{}^2$ where v_1 and v_2 are the initial and final velocities respectively.

Gain in potential energy $= \rho\Delta g H_2 - \rho\Delta g H_1$

(*Consider potential energy of a mass m raised through height h mgh*)

Thus, since the work done on the system = the gain in the energy of the system

So: $p_1\Delta - p_2\Delta = \dfrac{1}{2}\rho\Delta v_2{}^2 - \dfrac{1}{2}\rho\Delta v_1{}^2 + \rho\Delta g H_2 - \rho\Delta g H_1$

By sorting and cancelling terms on both sides of the equation, we obtain:

$$\dfrac{p_1}{\rho} + \dfrac{v_1{}^2}{2} + g H_1 = \dfrac{p_2}{\rho} + \dfrac{v_2{}^2}{2} + g H_2$$

Or simply:

$$\frac{p}{\rho} + \frac{v^2}{2} + gH = constant \qquad \textbf{(eq 8.13)}$$

Example 8.4: Consider $P_1 = 120,000$, Pa $v_1 = 6$ ms^{-1}, h$_1 = 0.0$ m, $v_2 = 7$ ms^{-1}, h$_2 = 2.0$ m, $\rho = 1000$ kg m^{-3}, and $g = 9.81$ ms^{-2}
Use Bernouilli's equation to find P_2.

$$\frac{P_1}{\rho} + \frac{v_1^2}{2} + gH_1 = \frac{P_2}{\rho} + \frac{v_2^2}{2} + gH_2$$

$$P_2 = \rho\left(\frac{P_1}{\rho} + \frac{v_1^2}{2} + gH_1 - \frac{v_2^2}{2} - gH_2\right)$$

So: $P_2 = 1000\left(\dfrac{P_1}{1000} + \dfrac{6^2}{2} + 9.81 \times 0.0 - \dfrac{7^2}{2} - 9.81 \times 2.00\right)$

$P_2 = 93880 \ Pa$

The working principle is that something makes the velocity of the fluid change, which produces a change in the pressure so that a difference $\Delta p = p_2 - p_1$ is created.

It can be shown for all these meters that the volume flow rate Q is related to Δp by the following formula:

$$Q = k(\Delta p)^{0.5} \qquad \textbf{(eq 8.14)}$$

where k is the meter constant.

Example 8.5: A Venturi meter has a meter constant of 0.008 m^4 N$^{-0.5}$ s^{-1}. Calculate the flow rate when $\Delta p = 210$ Pa (3 decimal places).

$Q = k(\Delta p)^{0.5}$

$Q = 0.008(210)^{0.5} = 0.116 \ m^3 s^{-1}$

8.8 Force sensors

The main types of force sensors are: mechanical, electrical strain gauge and hydraulic types.

8.8.1 Mechanical types

Mechanical types usually involve a spring, such as in a simple spring balance or bathroom scale. It is a basic mechanical principle that the deflection of a spring is directly proportional to the applied force, so if the movement is shown on a scale, the scale represents force.

Example 8.6: Consider a 1N weight, which extends a spring by 3cm. A second 1N weight is added, and the spring extends a further 3cm. What is the spring constant (2 decimal places)?

$$\Delta F = k\Delta x \qquad \textbf{(eq 8.15)}$$

$$1 = 0.03k$$

$$\text{So: } k = \frac{1}{0.03} = 33.33 \text{ Nm}^{-1}$$

8.8.2 Strain gauges

Strain gauges are used in many instruments that produce mechanical strain. They are used to measure the strain in a structure being stretched or compressed (see figure 8.11).

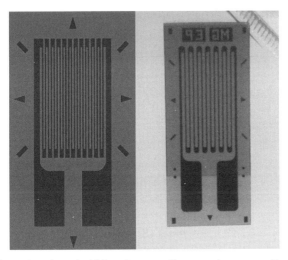

Figure 8.11 **a:** *Illustration of a typical foil strain gauge. The gauge is more sensitive to strain in the vertical direction than in the horizontal direction. Markings outside the active area help align the gauge during installation.* Figure 8.11 **b:** *A real strain gauge.*

The strain sensitive patterns may stretch due to tension, resulting in a narrower area with increased resistance. Compression, however, thickens the area, decreasing resistance. A strain gauge (or strain gage in the USA) is a device used to measure strain on any object and was invented almost simultaneously in America in 1938 by both the electrical engineer Edward E. Simmons and the mechanical engineer Arthur C. Ruge. The most common strain gauge consists of an insulating flexible backing, which supports a metallic foil pattern. The gauge is attached to the object by a suitable adhesive, such as cyanoacrylate. As the object shape is deformed, the foil is also deformed, causing its electrical resistance to change. This resistance change, usually measured using a Wheatstone bridge, is related to the strain by the quantity known as the *gauge factor*.

The strain gauge element is often a very thin wire formed into the shape shown. This produces a long wire all in one direction but on a small surface area. The element is often formed by etching a thin foil on a plastic backing. The completed element is then glued to the surface of the material or component that will be strained. The axis of the strain gauge is aligned with the direction of strain. When the component is stretched or compressed, the length of the resistance wire is changed. This produces a corresponding change in electrical resistance.

Let the length of the gauge be L and the change in length be ΔL.

The mechanical strain $\Delta = \dfrac{\Delta L}{R}$ (**eq 8.16**)

The resistance of the gauge is R (typically $120\,\Omega$) and the change in resistance ΔR.

Hence electrical strain $\xi = \dfrac{\Delta R}{R}$ (**eq 8.17**)

The electrical and mechanical strain are directly proportional and the constant relating them is called the gauge factor (typically 2).

Gauge factor = electrical strain/mechanical strain:

$$G = \frac{\xi}{\in} = \frac{L\Delta R}{R\Delta L} \qquad\qquad \textbf{(eq 8.18)}$$

Example 8.7: A strain gauge is glued with cyanoacrylate to a structure. It has a gauge factor of 2.2 and a resistance of $120.1\,\Omega$. The structure is stressed and the resistance changes to $120.15\,\Omega$. Calculate the strain and convert this into stress σ (4 significant figures). Take $E = 220\text{GPa}$.

$\Delta R = 120.15 - 120.1 = 0.05 \qquad \xi = \dfrac{0.05}{120.1} = 4.1632 \times 10^{-4}$

$$\epsilon = \frac{\xi}{G} = \frac{4.1632 \times 10^{-4}}{2.2} = 1.8923636 \times 10^{-4}$$

$$\sigma = \epsilon E \qquad \qquad \text{(eq 8.19)}$$

$\sigma = \epsilon E = 1.8923636 \times 10^{-4} \times 220\,G\,Pa = 41.63\,M\,Pa$

8.8.2.1 Further comments on the gauge factor

For metallic foil gauges, the gauge factor is usually a little over 2. For a single active gauge and three dummy resistors in a Wheatstone bridge configuration, the output from the bridge is: $v = \dfrac{BV \times G \times \epsilon}{4}$ where BV is the bridge excitation voltage.

Foil gauges typically have active areas of about 2–10 mm² in size. Strains up to at least 10 per cent can be measured.

8.8.3 Hydraulic types

Hydraulic types are often referred to as hydraulic load cells. The cell is a capsule filled with liquid. When the capsule is squeezed, the liquid becomes pressurised. The pressure represents the force and may be indicated with a calibrated pressure gauge. The capsule is often a short cylinder with a piston and the pressure produced is given by:$P = F/A$, where F is the force and A the cylinder area.

Pressure = force needed ÷ cylinder area

Example 8.8: What pressure is needed to develop 50,000 N of push force from a 10 cm diameter cylinder (1 decimal place)?

Force = 50,000N

Cylinder area = $\pi\,r^2$

Force needed ÷ cylinder area = $50,000 \div \pi\,r^2$

$= 50,000 \div \pi(0.1)^2 = 392.7 Nm^2$

8.9 Depth gauges

Depth gauges measure the depth of liquids and powder in tanks. They use a variety of principles and produce outputs in electrical and pneumatic forms. The type to use depends on the substance in the ship's hold or tank.

Here are a few examples: An ultrasonic system – e.g. reflected sound waves from the surface of a liquid – determines the depth from the time taken to receive the reflected sound. Electronic versions can use several electrical effects including fluid conductivity and capacitance changes. A pneumatic version will bubble air through a liquid and the pressure of the air is related to the depth. A simple pressure gauge attached to a tank indicates depth since depth is proportional to pressure.

There are, of course, many other sensors used in the maritime and port based environment, including inductive sensors, speed sensors, rheostats, tachometers and pH sensors, which – being of particular specialist interest – are not considered here, although readers will consider their own sensors and transducers to be more important!

Self-assessment questions

8.1   The cold resistance of a coil of wire is 22 Ω at 14 °C. It is heated to give a resistance of 26 Ω. Find its temperature rise, if the temperature coefficient of the resistance material is 0.0042 per degree centigrade (1 decimal place).

8.2   The cold resistance of a coil of wire is 24 Ω a t 10 °C. It is heated to give a resistance of 29 Ω. Find its temperature rise, if the temperature coefficient of the resistance material is 0.0045 per degree centigrade (1 decimal place).

8.3   The characteristic of a thermistor in terms of its resistance vs temperature generally follow the absolute value of its temperature described by the following equation:

$$\beta = \frac{\ln(\frac{R_{T_1}}{R_{T_2}})}{\left(\frac{1}{T_1} - \frac{1}{T_2}\right)} \text{ and}$$

R_{T_1} is the zero power resistance at absolute temperature T_1.

R_{T_2} is the zero power resistance at absolute temperature T_2, and

β is a constant whose value is determined by the composition of the thermistor material.

Find the β of a thermistor at $R_{30/60}$ whose values are $R_{30} = 8000$ ohms and $R_{60} = 2200$ ohms (4 significant figures).

8.4   For a steel-aluminum bimetallic strip, the following measurements were made: curvature $\delta = 0.020.5$m, $x = 0.190$m, the initial temperature is 23 °C.

Using: $\kappa = \dfrac{2\sin[tan^{-1}(\frac{\delta}{x})]}{\sqrt{(x^2 + d^2)}}$, find κ to 3 decimal places.

8.5   Find the mismatched strain $\Delta \epsilon$ (4 significant figures) using the expression below, and the values $h = 1$ mm and $E^* = 210 / 70$.

$$\Delta \epsilon = \frac{\kappa h(E^* + 14 + \frac{1}{E^*})}{12}$$

8.6   Using the expression $\Delta \epsilon = (a_A - a_B)\Delta T$, where $a_A = 1.5 \times 10^{-5}$ K^{-1} $a_B = 2.3 \times 10^{-5}$ K^{-1} and the value $\Delta \epsilon = 1.7 \times 10^{-3}$, find ΔT (2 decimal places).

8.7   Using the expression $Pa = P_0 + hg\rho$, find the atmospheric pressure if $P_0 = 90,000$ Pa, $= 9.81$ ms^{-2}, $h = 76.4$ mm Hg, and $\rho = 1000$ kg m^{-3}

8.8   Consider $P_1 = 100,000$, Pa $v_1 = 6$ ms^{-1}, $h_1 = 1.0$ m,

$v_2 = 9$ ms^{-1}, $h_2 = 2.0$ m, $\rho = 1000$ kg m^{-3}, and $g = 9.81$ ms^{-2}
Use Bernouilli's equation to find P_2.

$$\frac{P_1}{\rho} + \frac{v_1^2}{2} + gH_1 = \frac{P_2}{\rho} + \frac{v_2^2}{2} + gH_2$$

8.9   A Venturi meter has a meter constant of 0.015 m^4 N$^{-0.5}$ s^{-1}. Calculate the flow rate when $\Delta p = 200$ Pa (3 decimal places). Use:

$$Q = k(\Delta p)^{0.5}$$

8.10   Compare the advantages and disadvantages of different force sensors.

REFERENCES

[8.1] www.microchiptechno.com/ntc_thermistors.php

[8.2] *Longitude*, D Sobel (Fourth Estate, London, 1995, ISBN 978–1–8570–2502–6), p. 103.

APPENDIX 1

ANSWERS TO NUMERICAL QUESTIONS

Chapter 1

1.4. Base bandwidth $= f_H - f_L$

$= 1500 - 1000 = 1400$ Hz

1.6. Since $S = 12 + \dfrac{1}{R} - \dfrac{3}{R^2}$:

$$\frac{S}{N} = \frac{12 + \frac{1}{R} - \frac{3}{R^2}}{2} = 6$$

$$\frac{S}{N} = 6 + \frac{1}{2R} - \frac{3}{2R^2} = 6$$

$$(6 - 6) + \frac{1}{2R} - \frac{3}{2R^2} = 0$$

$$0 + \frac{1}{2R} - \frac{3}{2R^2} = 0$$

$$0 \times 2R^2 + \frac{1}{2R} \times 2R^2 - 3 = 0$$

$0 + R - 3 = 0$ Thus $R = 3$ km

1.7 Power Gain in dB $= 10 \log \left(\dfrac{P_{out}}{P_{in}} \right)$ or Gain $G = \dfrac{P_{out}}{P_{in}}$

So: $\dfrac{dB}{10} = \log \left(\dfrac{P_{out}}{P_{in}} \right)$

Taking inverse logarithms:

$$10^{\frac{dB}{10}} = \frac{P_{out}}{P_{in}}$$

Rearranging:

$$P_{out} = P_{in} \times 10^{\frac{dB}{10}}$$

$$P_{out} = 1 \times 10^{-3} \times 10^{\frac{35}{10}} = 3.16\,W$$

1.9   NIR light (optical fibre), microwave (radar waveguide), voice 3 kHz signal (twin pair), and 160 MHz (coaxial cable).

Chapter 2

2.4 $E(d) = E(0) \times e^{\frac{-3}{10}} = 0.74\,E(0)$ of the surface value, to 1 decimal place, a large fall-off in field strength.

2.6   OWF = 0.85 MUF = 0.85 × 1.2 MHz = 1.02 MHz. Any changes in the ionosphere or the range required will change the MUF.

2.8   A dipole antenna is in fact half a wavelength long and resonates at a frequency f calculated from this relationship: $f = \dfrac{c}{2L}$, where L is its length.

$$L = \frac{c}{2f} = \frac{3 \times 10^8}{2 \times 450 \times 10^6} = 0.33\,m$$

2.9   The tuned centre frequency is 7.3 MHz and the circuit has a Q-factor of 35. Using the equation:

$$B = \frac{f_c}{Q}$$

and substituting:

$$B = \frac{7.3 \times 10^6}{35} = 208.57\,kHz$$

Chapter 3

3.1   Electrical bandwidth $= f_H - f_L = 650\,kHz - 550\,kHz = 100\,kHz$

3.2   A dipole antenna is in fact half a wavelength long and resonates at a frequency *f* calculated from this relationship: $f = \dfrac{c}{2L}$>, where L is its length.

$$L = \frac{c}{2f} = \frac{3 \times 10^8}{2 \times 600 \times 10^6} = 0.25\,m$$

3.3   Consider the antenna structure shown in the figure below:

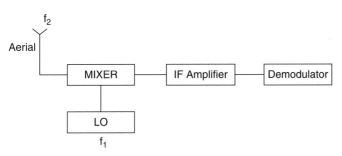

Figure: *Aerial arrangement: IF amplifier tuned to 500 kHz and bandwidth 20 kHz.*

(i)   Possibilities: Mixer outputs at the sum and difference frequencies.

Take firstly the differences with mostly $f_1 > f_2$

So let us look at the difference frequencies.
1.50 – 1.00 = 0.50 MHz
1.50 – 1.01 = 0.49 MHz
1.50 – 1.02 = 0.48 MHz
2.00 – 1.50 = 0.50 MHz

Now let us look at the sum frequencies.
1.50 + 1.00 = 2.50 MHz
1.50 + 1.01 = 2.51 MHz
1.50 + 1.02 = 2.52 MHz
1.50 + 2.00 = 3.50 MHz

(ii)   It is the 500 kHz *difference* between f_1 and f_2 that the IF is tuned to, ± 20 kHz so:

The IF amplifier will pass 500 kHz and 490 kHz.

(iii)   The circuit passes 1 MHz, 1.01 MHz and 1.02 MHz only. The 2 MHz frequency is rejected.

(iv)   Decreasing the IF bandwidth to 10 kHz, rejects all but the 1.00 MHz signal.

Chapter 4

4.1   The sequence of events can be seen below for any arbitrary 8-bit binary sequence:

Take a data sequence, say: 01011001

01011001	Plain text	(Data to be sent)
10110110	Keystream	(Encrypting code)
-----------------------	Modulo-2 addition	
11101111	Cypher text	(Transmitted secure data)

At the receiver an identical PRANG, using the same keymat, is used to generate a synchronised key stream. This is modulo-2 added to the received cypher text to recover the plain text.

11101111	Cypher text	(Secured received data)
10110110	Keystream	(Decrypting code)
---------------------	Modulus-2 addition	
01011001	Recovered plain text	

4.3   If there are 7000 samples sent per second, the bit rate will be $7000 \times 8 = 56$ k bits per second.

$$The\ minimum\ bandwidth\ B = \frac{bit\ rate}{2} = \frac{56000}{2} = 28\ kHz$$

4.5   If the highest frequency in an analogue video signal is 30 kHz, the necessary sampling frequency $= 2.2 \times f_H = 2.2 \times 30$ kHz $= 66000$ samples per second.

If there are 66000 samples sent per second, the bit rate will be 66000×16 $= 1.056$ M bits per second.

Chapter 5

5.2 Using the equation $R = \dfrac{ct}{2}$, R= $(3 \times 10^{8} \times 2.5 \times 10^{-3}) / 2 = 4.5 \times 10^{5}$m or 375 km

5.3 Using the for   mula:

$$MUR = \frac{c}{2PRF} = \frac{3 \times 10^{8}}{2 \times 1500} = 100\ km$$ The MDR is likely to be somewhat less than the MUR, so sufficient spare time exists to gather back more distant echoes before the next pulse is transmitted.

5.4 If the beam width is 1 degree, PRF = 2000 s^{-1} and the ARR = 10 RPM and using the equation:

$$N = \frac{a_H \times PRF}{6 \times ARR}$$

Substituting for the values given:

$$N = \frac{1 \times 2000}{6 \times 10} = 33$$

At least 33 pulses strike the target every revolution.

5.5 Using $Range\ Resolution = \dfrac{cT}{2} = \dfrac{3 \times 10^{8} \times 1 \times 10^{-6}}{2} = 150\ m$

5.6 Using the equation:

$$Doppler\ shift = 2f_{Transmitted} \times \frac{Relative\ Velocity}{c}$$

$$Doppler\ shift\ 50 = 10 \times 10^{9} \times \frac{Relative\ Velocity}{3 \times 10^{8}}$$

Thus relative velocity = 1.5 metres per second.

5.7 Gain = directional intensity/isotropic intensity = 100/5 = 20

dB Gain = 10 log (Gain) = 13.01 dB

Chapter 6

6.5 $V_o = GV_i = 35 \times 20 \times 10^{-3} = 0.7\,V$ saturated

$V_o = -G \log V_i$ *output in mV* $= -35 \times \log(20 \times 10^{-3}) = 59.46$ mV

(rectified) and thus is not saturated.

6.7   Ordinary mode range resolution $= \dfrac{c\tau}{2} = \dfrac{3 \times 10^8 \times 2.5 \times 10^{-6}}{2} = 375$ m

Pulse compression mode range resolution:

$$= \frac{c\tau}{2} = \frac{3 \times 10^8 \times 0.1 \times 10^{-6}}{2} = 15\,m$$

6.9   Using $\phi = 360\,k\sin\theta$

$\phi = 360 \times 0.5 \times \sin 30 = 90.0$ *degrees*

6.10   Synthesised SAR aperture $= v \times T = 210 \times 55 = 11.55$ km

For 20 GHz (1.5cm) waves, a 11.55 km antenna gives a $7.8 \times 10^{-5}\,°$ beam width from the beam width equation: $a_H = \dfrac{60\,\lambda}{D} = \dfrac{60 \times 0.015}{11550} = 7.8 \times 10^{-5}\,°$

Chapter 7

7.1 $E = hf = \dfrac{6.6\ 10^{-34}\ 3\ 10^8}{750\ 10^{-9}} = \dfrac{2.648\ 10^{-19}}{1.6\ 10^{-19}} = 1.655\ eV$

So kinetic energy $= 1.655 - 1.5 = 0.155$ eV

7.2   Using $\delta = \dfrac{Number\ of\ secondary\ electrons}{Number\ of\ incident\ electrons}$ $\delta = \dfrac{14}{5} = 2.8$

7.4 $E_n - E_{n-1} = 2.9$ eV $- 1.77$ eV $= 1.15$ eV

$$f = \frac{1.15\ 1.6\ 10^{-19}}{6.62\ 10^{-19}} = 2.779\ 10^{14} = 2.8\ 10^{14}\ Hz$$

7.6   Using: $I = \varepsilon\,\sigma\,T^4 = 0.27 \times 5.67 \times 10^{-8} \times (273.15 + 17)^4 = 108.5$ W

7.7 $\varepsilon = \dfrac{92}{100} = 0.92$

An aircraft's exhaust funnel has a uniform surface temperature of 250 °C. What will be the peak emission wavelength observed (2 significant figures).

$$\lambda_{peak} = \frac{2900}{(250 + 273.15)} = \lambda_{peak} = \frac{2900}{523.15} = 5.5 \; microns$$

7.8   Using the equation $I = \varepsilon \, \sigma \, T^4 \, W$

So:

$I_T = \varepsilon \, \sigma \, T^4$, and

$I_{1.15T} = \varepsilon \, \sigma \, (1.15T)^4 = 1.75 \, \varepsilon \, \sigma \, T^4$

Thus $\dfrac{I_{1.1T}}{I_T} = 1.75$, a 75 per cent increase in radiated intensity for a 15 per cent increase in surface temperature.

7.9   At thermal crossover:

$I_{Bridge} = \varepsilon_{Bridge}\sigma T_{Bridge}{}^4 = I_{River} = \varepsilon_{River}\sigma T_{River}{}^4$

$\varepsilon_{Bridge}\sigma T_{Bridge}{}^4 = \varepsilon_{River}\sigma T_{River}{}^4$

$\varepsilon_{Bridge}T_{Bridge}{}^4 = \varepsilon_{River}T_{River}{}^4$

$T_{Bridge}{}^4 = \dfrac{\varepsilon_{River}T_{River}{}^4}{\varepsilon_{Bridge}}$

$T_{Bridge} = \sqrt[4]{\dfrac{\varepsilon_{River}}{\varepsilon_{Bridge}}} \, T_{River}$

$T_{Bridge} = \sqrt[4]{\dfrac{0.92}{0.85}} \, (273.15 + 4) = 282.5 \; °C$

7.10   For a source radiating into space isotropically:

$$I = \frac{P}{4\pi R^2}$$

Thermally: $I = \dfrac{P}{A} = \varepsilon \, \sigma \, T^4$

Combining both of these equations by substitution for P:

$$I = \frac{\varepsilon A \sigma T^4}{4 \pi R^2}$$

Comparing two intensities for different emissivity, temperature and range conditions (assuming the emitting area A remains constant).

$$\frac{I_2}{I_1} = \frac{4\pi R_1^2 \varepsilon_2 A \sigma T_2^4}{4\pi R_2^2 \varepsilon_1 A \sigma T_1^4} = \frac{\varepsilon_2 T_2^4 R_1^2}{\varepsilon_1 T_1^4 R_2^2}$$

$$I_2 = I_1 \frac{\varepsilon_2 T_2{}^4 R_1{}^2}{\varepsilon_1 T_1{}^4 R_2{}^2}$$

$$I_2 = I_1 \frac{0.5 \, \varepsilon_1 (2T_1)^4 R_1{}^2}{\varepsilon_1 T_1{}^4 R_2{}^2}$$

$$I_2 = I_1 \frac{0.5 \times 16 \, R_1{}^2}{R_2{}^2} = I_1 \frac{8 \, R_1{}^2}{R_2{}^2}$$

Chapter 8

8.1:   The cold resistance of a coil of wire is 22 Ω at 14°C. It is heated to give a resistance of 26 Ω. Find its temperature rise, if the temperature coefficient of the resistance material is 0.0042 per degree centigrade (1 decimal place).

Thus: using $R = R_0(1 + a \, T)$

for the two temperatures T_1 and T_2.

$$\frac{R_2}{R_1} = \frac{R_0(1 + a \, T_2)}{R_0(1 + aT_1)}$$

Thus:

$$\frac{26}{22} = \frac{R_0(1 + 0.0042 \, T_2)}{R_0(1 + 0.0042 \, T_1)}$$

$$\frac{26}{22} = \frac{R_0(1 + 0.0042 \, T_2)}{R_0(1 + 0.0042 \times 14)}$$

Hence: $1.18181 \times 1.0588 = 1 + 0.0042 \, T_2$

So $T_2 = 59.8 \, °C$

And the temperature rise is $59.8 - 14 = 45.8 °C$

8.2:   Thus: using $R = R_0(1 + aT)$

for the two temperatures T_1 and T_2.

$$\frac{R_2}{R_1} = \frac{R_0(1 + a\, T_2)}{R_0(1 + a\, T_1)}$$

Thus:

$$\frac{29}{24} = \frac{R_0(1 + 0.0045\, T_2)}{R_0(1 + 0.0045\, T_1)}$$

$$\frac{29}{24} = \frac{R_0(1 + 0.0045\, T_2)}{R_0(1 + 0.0045 \times 10)}$$

Hence: $1.2627083 \times 1.045 = 1 + 0.0045\, T_2$

So $T_2 = 58.4\ ^\circ C$

And the temperature rise is $58.4 - 10 = 48.4\ ^\circ C$

8.3: $T_1 = 273.15 + 30 = 303.15\text{K}$ $T2 = 273.15 + 60 = 333.15\text{K}$

$$\beta = \frac{\ln\left(\frac{8000}{2200}\right)}{\left(\frac{1}{303.15} - \frac{1}{333.15}\right)} \quad \beta = 4346$$

8.4: $\kappa = \dfrac{2\sin[tan^{-1}(\frac{0.0205}{0.190})]}{\sqrt{(0.0205^2 + 0.190^2)}} = 1.123$

8.5: $\Delta \in = \dfrac{\kappa h(E^* + 14 + \frac{1}{E^*})}{12}$

$$\frac{I_2}{I_1} = \frac{4\pi\, R_1^2 \varepsilon_2 A\, \sigma\, T_2^4}{4\pi\, R_2^2 \varepsilon_1 A\, \sigma\, T_1^4} = \frac{\varepsilon_2 T_2^4 R_1^2}{\varepsilon_1 T_1^4 R_2^2}$$

8.6: $\Delta T = \dfrac{\Delta \varepsilon}{(a_A - a_B)} = \dfrac{1.7 \times 10^{-3}}{(1.5 \times 10^{-5} - 2.3 \times 10^{-5})} = -212.5\ kelvin$

8.7: $Pa = P_0 + hg\rho$

$Pa = 90{,}000 + 76.4 \times 10^{-3} \times 9.81 \times 1000 = 90749.48 Pa$

8.8:

$$\frac{P_1}{\rho} + \frac{v_1^2}{2} + gH_1 = \frac{P_2}{\rho} + \frac{v_2^2}{2} + gH_2$$

$$P_2 = \rho\left(\frac{P_1}{\rho} + \frac{v_1^2}{2} + gH_1 - \frac{v_2^2}{2} - gH_2\right)$$

So: $P_2 = 1000\ (\frac{100000}{1000} + \frac{6^2}{2} + 9.81 \times 1.0 - \frac{9^2}{2} - 9.81 \times 2.00)$

$$P_2 = 83690\ Pa$$

8.9:

$$Q = 0.015(200)^{0.5} = 0.122\ m^3 s^{-1}$$

GLOSSARY

Amplitude modulation The modulation of a wave by varying its amplitude, especially as a means of broadcasting an audio signal by combining it with a radio carrier wave.

Amplitude Shift Keying (ASK) A simple method of digital keying, similar in principle to analogue amplitude modulation.

Analogue signal A signal that varies continuously between preset limits. A sinusoid or sine wave is an example of an analogue signal.

Antenna gain (G) A measure of the directivity of an antenna, expressed as a ratio of the intensity of the directional antenna's main beam to the intensity of an equally powerful isotropic radiator at the same range.

Aperture The effective radiating or receiving area of a radar antenna.

Array A number of radar sources or receivers used in conjunction as an antenna. The sources can be arranged in one dimension (a linear array) or in two dimensions (a planar array).

Asynchronous signalling Signalling that uses 'start' and 'stop' elements to denote the beginning and end of each character code. This means that the equipment required at the transmitting and receiving stations does not need very accurate clocking as is required in synchronous signalling.

Atmospheric noise Radio noise caused by natural atmospheric processes, primarily lightning discharges in thunderstorms. On a worldwide scale, there are about 40 lightning flashes per second – approximately 3.5 million lightning discharges per day.

Automatic Request for Repetition (ARQ) An EDAC system that allows the receiving station to detect data errors and request the transmitting station to retransmit the affected data. The number of parity bits required for ARQ is generally smaller than for FEC.

Baud speed The number of carrier condition changes occurring in one second. These changes can be in the carrier amplitude or angle (frequency or phase) and the baud speed governs the bandwidth requirement of the signal.

Bearer A medium for conveying a signal from one place to another. All bearers are suitable for all types of signalling if appropriate bearer conditioning is carried out. Examples of bearers are: twisted pair, coaxial cables, radio frequencies, waveguides and optical fibres.

Binary digit (BIT/bit) Bit represents one piece of digital information. In a digital communications system, the duration of all bits is the same.

Bit rate The number of bits or the amount of data that are processed over a certain amount of time. In audio, this usually means kilobits per second. For example, the music you buy on iTunes is 256 kilobits per second, meaning there are 256 kilobits of data stored in every second of a song.

Bluetooth Bluetooth and Bluetooth Low Energy (BLE) are wireless technologies used to transfer data over short distances. The technology is frequently used in small devices that connect to users' phones and tablets. For instance, the technology is used in many speaker systems.

Boresight The direction at right angles to the surface of an array or the axis of symmetry of a parabolic dish. The boresight is used as a reference for the measurement of beam widths.

Bowman A family of tactical digital radios used by the UK armed forces.

Broadband antenna An antenna with a response over a wide range of frequencies with a low 'Q'.

Burst transmission A method of sending a large amount of data in a very short period of time, often used in military communications traffic.

Byte A larger and more convenient quantity for representing digital data than a bit (1 byte = 8 bits).

Capacity Channel capacity is determined by the amount of frequency bandwidth that channel has available. Every communications channel, whether optical fibre or not, can support or carry at least one independent set of information. However, the capacity of some bearers, such as optical fibre, is much greater than the capacity required to carry just one set of information, whether audio or video. Such bearers can thus support multiple channels.

Carrier A signal at the required transmission frequency modulated by the baseband to give the required signal format. The carrier may be Amplitude Modulated (AM), Frequency Modulated (FM) or Phase Modulated (PM) by the baseband.

Clutter Radar clutter consists of unwanted echoes from small multiple reflectors, such as raindrops, which compete with target echoes.

Communications satellite A device used to extend the radio horizon for space wave communications by providing a high altitude relay. Most communications satellites are in synchronous (geostationary) orbit around the equator.

Continuous Wave (CW) radars Radars that emit a continuous stream of electromagnetic waves. Particularly useful for measuring target velocities by the Doppler effect.

Critical frequency The highest frequency at which a vertically transmitted signal (transmitted at zero angle of incidence) will undergo total internal reflection from the ionosphere. The critical frequency will change depending on the state of the ionosphere.

Cryptography The process of providing security to a signal by encrypting the data prior to transmission and decrypting it after reception. The encryption/decryption process can be online (synchronous, real-time) or offline (asynchronous, non-real-time).

Cypher text In cryptography, a cypher text is the result of encryption performed on plain text using an algorithm, called a cypher.

Data rate The rate at which data is transmitted, measured in bits per second (bps). The data rate is equal to the number of carrier condition changes per second (baud speed) times the number of bits of information carried by each carrier condition. For binary signals, data rate and baud speed are identical.

Differentiation A radar anti-clutter technique that displays only the leading edge of targets to prevent receiver saturation.

Digital signal A signal containing instantaneous transitions between two predetermined states. A square wave is an example of a digital signal.

Dipole An antenna having a length equal to half the radiating wavelength.

Double Side Band (Supressed Carrier) DSB(SC). An AM broadcast method.

Duplexer A device that prevents high power transmitted pulses from entering the receiver or low power received echoes from entering the transmitter. It consists of a TR switch and an anti-TR switch.

Electron emission The emission of electrons by materials caused by heat (thermionic emission), incident radiation (photoelectric emission) or incident particles (secondary emission).

Emergency Position Indicating Radio Beacon (EPIRB) A device for transmitting a distress alert so the distress position can be found.

Encoding Plain text digital samples converted into a series of ones and zeroes.

Error Detection and Correction (EDAC) Communication technique that uses parity bits to allow a receiving station to detect and correct errors in received data. EDAC requires more bits to be sent than just the data and therefore reduces the overall rate at which information is transferred.

Filter Any device, either analogue or digital, in communications or sensing, that allows the selection of a particular signal against other unwanted signals and noise.

Forward Error Correction (FEC) FEC is a form of EDAC that allows the receiving station to detect and correct data errors without asking the transmitting station to send the data again. FEC systems require more parity bits than ARQ systems, but are useful on one-way (broadcast) or unrepeatable transmissions, e.g. transmissions from space probes and satellites.

Frequency The number of cycles, vibrations or oscillations per second of a movement or event repeating per unit time.

Frequency hopping A spread spectrum technique that provides security, anti-interference and anti-noise capability to a communications system by changing the transmission frequency using a pseudo-random pattern. The systems are known as slow or fast hopping systems, depending on the number of frequency changes performed per minute.

Frequency Shift Keying (FSK) A frequency modulation scheme in which digital information is transmitted through discrete frequency changes of a carrier signal. The technology is used for communication systems such as amateur radio, caller ID and emergency broadcasts. The simplest FSK is Binary FSK (BFSK).

Frequency spectrum The electromagnetic spectrum is the collective term for all possible frequencies of electromagnetic radiation.

Galactic noise This is produced from non-terrestrial sources.

Global Maritime Distress Safety System (GMDSS) A maritime communications system for all vessels. It follows internationally agreed safety procedures and specifies the types of equipment used to increase safety, in order to make it easier to rescue ships, boats and aircraft.

Gridded network This sort of network has built-in bearer redundancy by providing many interconnections between switches. A gridded network provides increased survivability to the network by introducing graceful degradation instead of catastrophic failure.

Ground wave A propagation mode of electromagnetic radiation that consists of direct waves, ground reflected waves and surface waves up to the visible horizon. Beyond the horizon, only the surface diffracted wave remains, which is only significant at HF or below.

H3E single sideband A single sideband voice, full carrier transmission method used on SSB radio when receiving broadcast stations, e.g. the BBC World Service.

Image Intensifier (II) A device that uses photoelectric and secondary emission effects to amplify the incident light level.

Integration The summing of radar pictures on a pulse by pulse basis. Integration improves a radar's Signal/Noise (S/N) ratio.

Internet The internet is the global system of interconnected mainframe, personal and wireless computer networks that use the internet protocol suite (TCP/IP) to link billions of devices worldwide.

Ionosphere A region of the atmosphere, typically between 50 and 400 km in height, which contains layers of ionised gases. The layers are known as the E, F_1 and F_2 and through the processes of refraction and total internal reflection produce sky waves for MF and HF frequencies.

Ionospheric scatter Radiation can undergo scattering from precipitation, from frequent rain or spray particles suspended in the atmosphere or from patches of abnormally high ionisation created by meteoric paths.

Keying Keying techniques are those used to modulate a carrier wave with a digital signal. They can alter the carrier's Amplitude (ASK), Frequency (FSK) or Phase (PSK) and are similar to analogue: AM, FM and PM.

Key stream The Pseudo Random Number Sequence (PRNS) generated by the Pseudo RAndom Number Generator (PRANG) is referred to as the 'key stream'.

LASER Acronym for Light Amplification by the Stimulated Emission of Radiation. Lasers convert electrical energy into tight, coherent beams of visible or IR radiation. Lasers can be either continuous or pulsed. They are mainly used as range finders (laser radars), target designators or ASK transmitters.

Linear network A collection of switches where each switch is only connected to two other switches, with no branching or other interconnections. Signal traffic flows linearly along the network and each switch is an access switch for local users.

Logarithmic amplification An anti-clutter technique that prevents saturation of a radar receiver. The gain of the receiver is made dependent on the logarithm of the input power.

Lowest Usable Frequency (LUF) The lowest usable frequency to give sky wave communications between a given transmitter and receiver. This lower limit is caused by attenuation loss of the signal in the D-region of the ionosphere and can be altered by changing the transmitted power.

Marine Very High Frequency (VHF) radio This radio uses the radio frequency range between 156.0 and162.025 MHz, called the VHF maritime mobile band. Marine radio equipment is installed on all large ships and most seagoing small craft. VHF is Line Of Sight (LOS).

Master Oscillator Power Amplifier (MOPA) transmitter A radar transmitter that uses a stable oscillator followed by a modulator and power amplifier to produce high powered electromagnetic wave pulses.

Master trigger unit A device that co-ordinates the firing of all a ship's radars by synchronising their PRFs. They also provide blanking pulses for all vulnerable non-radar receivers.

Maximum Detection Range (MDR) The maximum range at which a radar can detect a target. The MDR is determined in part by the duration of the pulse – the longer the pulse, the more energy there is in the pulse and the longer the MDR.

Maximum Unambiguous Range (MUR) The limit of unambiguous ranging of a radar. The MUR should be longer than the MDR to prevent ambiguous ranging. The MUR is determined by the PRF of a radar.

Maximum Usable Frequency (MUF) The maximum usable frequency to give sky wave communications between a given transmitter and receiver. The MUF depends on the state of the ionosphere and the geographical locations of both the transmitter and receiver and cannot be changed by changing the transmitted power. The MUF is the highest frequency that keeps the receiver in the reception zone.

Mixing This process involves the multiplication of two or more signals to give components at the sum and difference of the signal frequencies. Mixers are used as modulators, demodulators and frequency changers in transmitters and receivers.

Modem Modem is short for modulator/demodulator and is a device that modulates an analogue carrier with digital data so that it can be sent over a twisted pair or other telephone line. At the receiver, the modem demodulates the received signal to recover the digital data. Modems use one of the keying techniques to achieve this.

Modulation index The modulation index (or modulation depth) of a modulation scheme describes by how much the modulated variable of the carrier signal varies around its unmodulated level.

Monopole transmitter A transmitter that has an aerial that is ¼ of the radio antenna wavelength.

Moving target indication A radar anti-clutter technique that provides sub-clutter visibility. It works by discriminating between targets and clutter sources by virtue of their difference in velocity.

Multi-channelling A method involving processes for transmitting several independent sets of information over a wideband communications channel.

Multiple access Multi-channelling is achieved by allocating frequency bands (Frequency Division Multiple Access, FDMA), time slots (Time Division Multiple Access, TDMA) or codes (Code Division Multiple Access, CDMA) to various users.

Multiplexing This is achieved by the controller of the channel by allocating frequency bands (Frequency Division Multiplexing, FDM) or time slots (Time Division Multiplexing, TDM) to different signals. FDM can be used with both analogue and digital signals, but TDM can only be used with digital signals.

Narrowband antenna An antenna with a very sharp resonance and response over a narrow range of frequencies, having a high 'Q'.

Network Any system that allows 'networked' devices, sensors or transducers to exchange data with each other along network links. The connections between devices is established using either cable media or wireless media.

Noise Any unwanted signal that interferes with a wanted signal. Noise can be external, originating outside the receiver, or internal, originating inside the receiver. All signals must be read against a noise background. Noise can be generated naturally or be man-made, either accidentally (by interference) or deliberately (by jamming).

Optical fibre A flexible, transparent fibre made by drawing glass (silica) or plastic to a diameter slightly thicker than that of a human hair.

Parity bits Bits other than data bits that are transmitted for EDAC purposes.

Phased arrays An array of sources and/or receivers used as an antenna. Arrays may be passive, where a single remote power source supplies all the array elements, or active, where each element has its own power source. An adaptive array is one that can automatically react to events and can continue to function in a multiple jamming environment. Arrays are now commonly used in communications antennas, radar systems, and in optical and infrared focal plane arrays.

Phase Shift Keying (PSK) A digital modulation scheme that conveys data by changing (modulating) the phase of a reference signal (the carrier wave).

Plain text Data text that is not specially formatted, or written in code.

Pointing radars Radars that point at a target and track it. They are mainly used as tracking or fire control radars and have a narrow beam width.

PRF stagger The use of two or more PRFs to remove ambiguous ranges in high PRF radars. Ranges greater than the MUR are discarded by the radar and unambiguous ranges are obtained.

PRF switching A similar technique to PRF stagger except that ranges above the MUR are calculated rather than being discarded. Used in very high PRF tracking radars.

Pulse Repetition Interval (PRI) PRI is the time interval between the transmission of successive radar pulses.

Pulse duration The pulse duration of a radar (τ) is the time for which each pulse is transmitted.

Pulse Repetition Frequency (PRF) The rate at which the radar produces pulses. The units are Hz or pulses per second.

Pulsed radars Radars that transmit high power pulses of electromagnetic radiation and obtain the range by measurement of the elapsed time.

Radar Cross Section (RCS or σ) The effective 'echoing' area that a target presents to a radar. This has three components: the physical size of the target, the reflectivity of the target (material) and the directivity of the target (shape).

Radio receiver A device that selects a wanted signal from any unwanted signals, and retrieves the modulation information contained in the detected signal.

Radome A solid enclosure for radars to protect them from the effects of weather, particularly wind, icing and precipitation.

Reception zone The area in sky wave communications between the first returning sky wave ('skip distance') and the last readable sky wave. Within the reception zone there is continuous sky wave reception.

Repeater A device used on bearers to amplify a transmitted signal and to restore the S/N ratio of the signal. Repeaters can be processing (take the signal back to the baseband) or non-processing (amplification of the signal directly).

Resolution The ability of a radar to resolve targets in range (range resolution), bearing (bearing resolution) or elevation (elevation resolution). The resolution is controlled by the pulse duration, the horizontal beam width and the vertical beam width of the radar.

Sensitivity The more sensitive a radar receiver, the weaker the signal that can be detected against a background of noise.

Silent zone In sky wave communications, the area where receivers in the silent zone will not receive any signal from the transmitter. The silent zone extends from the end of the ground wave to the first returning sky wave.

Single Side Band An AM modulation method that eliminates either the upper or lower sideband, thus reducing signal bandwidth.

Single Side Band (Suppressed Carrier) SSB(SC) AM method that removes one sideband and the carrier wave as well.

Skip distance In sky wave communications, the distance from the transmitter to the point at which the first returning sky wave returns to Earth. The skip distance is controlled by the ionosphere and the transmitted frequency – the higher the transmitted frequency, the longer the skip distance.

Sky wave A propagation mode involving refraction and total internal reflection in the ionosphere to return frequencies in the MF and HF bands back to Earth. Sky waves are the main long range communication method in these bands.

Space wave A propagation mode involving electromagnetic waves propagating as either direct or ground reflected waves. The range of the space wave is limited for surface communications by the height of the transmitter and receiver antennas.

Spread spectrum This covers a range of techniques used to provide security and anti-interference capability to signals. Spread spectrum can be achieved as either Direct Sequence Spread Spectrum (DSSS), which spreads the signal over a wide bandwidth, or frequency hopping, which moves a narrowband signal pseudo-randomly within a wideband channel.

Star network A network that consists of just one central switch, hub or computer, which enables the transmission of messages between different nodes. The central node is connected to all other nodes and provides a common connection point for all nodes through the hub.

Surface wave A propagation mode that involves electromagnetic waves diffracted around the surface of the Earth. It is most useful for long range communications at LF and VLF, but can be used to provide Limited Range of Intercept signals using groundwaves.

Surveillance radars Radars used to give all round coverage and to detect all targets in their hemispherical coverage.

Swept gain A radar anti-clutter technique used to reduce the effects of close range sea clutter. The gain of the receiver is reduced at the beginning of every cycle and then returns to its maximum value during the cycle. This prevents sea clutter from saturating the receiver and allows strong targets to be seen. Swept gain does not provide sub-clutter visibility of weak targets.

Synchronous satellite A satellite orbit in which the satellite orbits the equator once every 24 hours. The satellite therefore appears to be stationary over one point on the equator, called a geostationary orbit. The orbital height of the satellite is about 22,500 miles or 35,000 km.

Synchronous signalling A real-time signalling method in which timing information is transmitted separately along with the data. Synchronous signalling can be applied to data exchange systems that are synchronised once every 24 hours and kept synchronised by accurate clocks at the transmitter and receiver stations, e.g. in a maritime broadcast.

Tapered illumination A technique for reducing radar and communications antenna side lobes by reducing the power supplied to the outer parts of an antenna, especially with phased array radar.

Telegraphy The long distance transmission of textual or symbolic (as opposed to verbal or audio) messages without the physical exchange of an object bearing the message.

Thermal imager A device using Infra-Red (IR) or heat detectors to provide an infrared (heat) picture of a target or area.

Threshold detection The returned echoes and noise in a radar are compared to a threshold voltage and all peaks above the threshold are registered as targets and displayed. Setting the threshold too high will cause weak targets to be discarded (missed contacts) and setting it too low will cause noise spikes to be registered as targets (false alarms).

Track While Scan (TWS) The use of range, angle and velocity gates to enable a surveillance radar to measure a target's range, bearing and speed. These measurements are used to calculate a target's motion and to drive a tracking window on the display.

Trigger unit The internal radar clock providing the pulses to control the PRF.

Troposphere A region of the atmosphere extending from the surface of the Earth to the tropopause at a height of about 10 km. The troposphere has a negative temperature gradient and is characterised by large masses of air undergoing continual convective motion. In communication terms it is largely homogenous, but causes some bending of electromagnetic waves, extending the radio horizon to about $^4/_3$ of the visual horizon.

Tropospheric scatter Tropospheric scattering is caused by continuous atmospheric events. The scattered signal is fairly independent of atmospheric conditions, giving reliable communications.

Twisted pair Twisted pair cabling is a type of wiring in which two conductors of a single circuit are twisted together for the purposes of cancelling out ElectroMagnetic Interference (EMI) from external sources; for instance, electromagnetic radiation from unshielded twisted pair (UTP) cables, and crosstalk between neighbouring pairs.

Voltage Control Oscillator (VCO) A device that outputs a sinusoid of a frequency that is a function of the input voltage.

Waveguide A structure that guides waves, such as electromagnetic waves or sound waves. They enable a signal to propagate with minimal loss of energy by restricting expansion to one or two dimensions.

Wireless technology Any technology that allows devices to transfer information between them without direct contact, usually achieved by optical, Radio Frequency (RF), or microwave frequency methods.

Index

Note: Page numbers followed by 'f' and 't' refer to figures and tables, respectively.